Dimensions Math®
Teacher's Guide KA

Authors and Reviewers

Cassandra Turner

Elizabeth Curran

Allison Coates

Tricia Salerno

Pearly Yuen

Jenny Kempe

Singapore Math Inc.

Published by Singapore Math Inc.

19535 SW 129th Avenue
Tualatin, OR 97062
www.singaporemath.com

Dimensions Math® Teacher's Guide Kindergarten A
ISBN 978-1-947226-30-2

First published 2018
Reprinted 2019, 2020

Printed in China

Acknowledgments

Editing by the Singapore Math Inc. team.
Design and illustration by Cameron Wray with Carli Fronius.

Contents

Chapter		Lesson	Page

Chapter 1
Match, Sort, and Classify

		Teaching Notes	1
		Chapter Opener	5
	1	Left and Right	6
	2	Same and Similar	8
	3	Look for One That is Different	10
	4	How Does it Feel?	12
	5	Match the Things That Go Together	14
	6	Sort	16
	7	Practice	17
		Workbook Pages	19

Chapter 2
Numbers to 5

		Teaching Notes	23
		Chapter Opener	27
	1	Count to 5	28
	2	Count Things Up to 5	31
	3	Recognize the Numbers 1 to 3	34
	4	Recognize the Numbers 4 and 5	37
	5	Count and Match	39
	6	Write the Numbers 1 and 2	41
	7	Write the Number 3	44
	8	Write the Number 4	46
	9	Trace and Write 1 to 5	48
	10	Zero	50
	11	Picture Graphs	53
	12	Practice	55
		Workbook Pages	57

Chapter		Lesson	Page

Chapter 3
Numbers to 10

		Page
	Teaching Notes	65
	Chapter Opener	69
1	Count 1 to 10	70
2	Count Up to 7 Things	71
3	Count Up to 9 Things	73
4	Counting Up to 10 Things — Part 1	76
5	Counting Up to 10 Things — Part 2	79
6	Recognize the Numbers 6 to 10	80
7	Write the Numbers 6 and 7	82
8	Write the Numbers 8, 9, and 10	84
9	Write the Numbers 6 to 10	86
10	Count and Write the Numbers 1 to 10	88
11	Ordinal Positions	90
12	One More Than	93
13	Practice	95
	Workbook Pages	97

Chapter 4
Shapes and Solids

		Page
	Teaching Notes	109
	Chapter Opener	113
1	Curved or Flat	114
2	Solid Shapes	116
3	Closed Shapes	118
4	Rectangles	119
5	Squares	121
6	Circles and Triangles	123
7	Where is It?	126
8	Hexagons	128
9	Sizes and Shapes	130
10	Combine Shapes	132
11	Graphs	134
12	Practice	136
	Workbook Pages	138

Teacher's Guide KA

Chapter		Lesson	Page

Chapter 5
Compare Height, Length, Weight, and Capacity

		Lesson	Page
		Teaching Notes	145
		Chapter Opener	149
	1	Comparing Height	150
	2	Comparing Length	153
	3	Height and Length — Part 1	156
	4	Height and Length — Part 2	158
	5	Weight — Part 1	160
	6	Weight — Part 2	162
	7	Weight — Part 3	163
	8	Capacity — Part 1	165
	9	Capacity — Part 2	167
	10	Practice	169
		Workbook Pages	171

Chapter 6
Comparing Numbers Within 10

		Lesson	Page
		Teaching Notes	177
		Chapter Opener	181
	1	Same and More	182
	2	More and Fewer	184
	3	More and Less	187
	4	Practice — Part 1	190
	5	Practice — Part 2	192
		Workbook Pages	193

Resources

	Blackline Masters for KA	197

Blank

Dimensions Math® Curriculum

The **Dimensions Math®** series is a Pre-Kindergarten to Grade 5 series based on the pedagogy and methodology of math education in Singapore. The main goal of the **Dimensions Math®** series is to help students develop competence and confidence in mathematics.

The series follows the principles outlined in the Singapore Mathematics Framework below.

Pedagogical Approach and Methodology

- Through Concrete-Pictorial-Abstract development, students view the same concepts over time with increasing levels of abstraction.
- Thoughtful sequencing creates a sense of continuity. The content of each grade level builds on that of preceding grade levels. Similarly, lessons build on previous lessons within each grade.
- Group discussion of solution methods encourages expansive thinking.
- Interesting problems and activities provide varied opportunities to explore and apply skills.
- Hands-on tasks and sharing establish a culture of collaboration.
- Extra practice and extension activities encourage students to persevere through challenging problems.
- Variation in pictorial representation (number bonds, bar models, etc.) and concrete representation (straws, linking cubes, base ten blocks, discs, etc.) broaden student understanding.

Each topic is introduced, then thoughtfully developed through the use of a variety of learning experiences, problem solving, student discourse, and opportunities for mastery of skills. This combination of hands-on practice, in-depth exploration of topics, and mathematical variability in teaching methodology allows students to truly master mathematical concepts.

Singapore Mathematics Framework

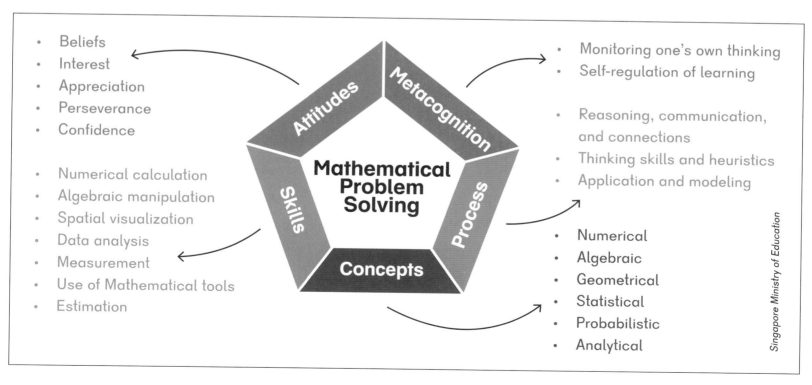

Singapore Ministry of Education

Dimensions Math® Program Materials

Textbooks

Textbooks are designed to help students build a solid foundation in mathematical thinking and efficient problem solving. Careful sequencing of topics, well-chosen problems, and simple graphics foster deep conceptual understanding and confidence. Mental math, problem solving, and correct computation are given balanced attention in all grades. As skills are mastered, students move to increasingly sophisticated concepts within and across grade levels.

Students work through the textbook lessons with the help of five friends: Emma, Alex, Sofia, Dion, and Mei. The characters appear throughout the series and help students develop metacognitive reasoning through questions, hints, and ideas.

A pencil icon ➤ at the end of the textbook lessons links to exercises in the workbooks.

Workbooks

Workbooks provide additional problems that range from basic to challenging. These allow students to independently review and practice the skills they have learned.

Teacher's Guides

Teacher's Guides include lesson plans, mathematical background, games, helpful suggestions, and comprehensive resources for daily lessons.

Tests

Tests contain differentiated assessments to systematically evaluate student progress.

Emma   Alex   Sofia   Dion   Mei

Online Resources

The following can be downloaded from dimensionsmath.com.

- **Blackline Masters** used for various hands-on tasks.

- **Letters Home** to be emailed or sent home with students for continued exploration. These outline what the student is learning in math class and offer suggestions for related activities at home. Reinforcement at home supports deep understanding of mathematical concepts.

- **Videos** of popular children songs used for singing activities.

- **Material Lists** for each chapter and lesson, so teachers and classroom helpers can prepare ahead of time.

- **Activities** that can done with students who need more practice or a greater challenge, organized by concept, chapter, and lesson.

- **Standards Alignments** for various states.

Using the Teacher's Guide

This guide is designed to assist in planning daily lessons, and should be considered a helping hand between the curriculum and the classroom. It provides introductory notes on mathematical content, key points, and suggestions for activities. It also includes ideas for differentiation within each lesson, and answers and solutions to textbook and workbook problems.

Each chapter of the guide begins with the following.

- ## Overview

  Includes objectives and suggested number of class periods for each chapter.

- ## Notes

  Highlights key learning points, provides background on math concepts, explains the purpose of certain activities, and helps teachers understand the flow of topics throughout the year.

- ## Materials

  Lists materials, manipulatives, and Blackline Masters used in the Explore and Learn sections of the guide. It also includes suggested storybooks. Blackline Masters can be found at dimensionsmath.com.

The guide goes through the Chapter Openers, Daily Lessons, and Practices of each chapter in the following general format.

- ## Explore

  Introduces students to math concepts through hands-on activities that can be done in small or whole group formats.

- ## Learn

  Summarizes the main concepts of the lesson, including exercises from Look and Talk pages. Small and whole group activities are often utilized in this section.

● <u>Whole Group</u> & <u>Small Group Activities</u>

Allows students to practice concepts through hands-on tasks and games, including suggestions for outdoor play (most of which can be modified for a gymnasium or classroom).

Level of difficulty in the games and activities are denoted by the following symbols.

- ● Foundational activities
- ▲ On-level activities
- ★ Challenge or extension activities

● <u>Extend</u>

This expands on <u>**Explore**</u>, <u>**Learn**</u>, and <u>**Activities**</u> and provides opportunities for students to deepen their understanding and build confidence.

Discussion is a critical component of each lesson. Have students share their ideas with a partner, small group, or the class as often as possible. As each classroom is different, this guide does not anticipate all situations. Teachers are encouraged to elicit higher level thinking and discussion through questions like these:

- Why? How do you know?
- Can you explain that?
- Can you draw a picture of that?
- Does your answer make sense? How do you know?
- How is this task like the one we did before? How is it different?
- What did you learn before that can help you to solve this problem?
- What is alike and what is different about this?
- Can you solve that a different way?
- How do you know it's true?
- Can you restate or say in your own words what your classmate shared?

Lesson structures and activities do not have to conform exactly to what is shown in the guide. Teachers are encouraged to exercise their discretion in using this material in a way that best suits their classes.

Dimensions Math® Scope & Sequence

PKA

Chapter 1
Match, Sort, and Classify

Red and Blue
Yellow and Green
Color Review
Soft and Hard
Rough, Bumpy, and Smooth
Sticky and Grainy
Size — Part 1
Size — Part 2
Sort Into Two Groups
Practice

Chapter 2
Compare Objects

Big and Small
Long and Short
Tall and Short
Heavy and Light
Practice

Chapter 3
Patterns

Movement Patterns
Sound Patterns
Create Patterns
Practice

Chapter 4
Numbers to 5 — Part 1

Count 1 to 5 — Part 1
Count 1 to 5 — Part 2
Count Back

Count On and Back
Count 1 Object
Count 2 Objects
Count Up to 3 Objects
Count Up to 4 Objects
Count Up to 5 Objects
How Many? — Part 1
How Many? — Part 2
How Many Now? — Part 1
How Many Now? — Part 2
Practice

Chapter 5
Numbers to 5 — Part 2

1, 2, 3
1, 2, 3, 4, 5 — Part 1
1, 2, 3, 4, 5 — Part 2
How Many? — Part 1
How Many? — Part 2
How Many Do You See?
How Many Do You See Now?
Practice

Chapter 6
Numbers to 10 — Part 1

0
Count to 10 — Part 1
Count to 10 — Part 2
Count Back
Order Numbers
Count Up to 6 Objects
Count Up to 7 Objects
Count Up to 8 Objects
Count Up to 9 Objects
Count Up to 10 Objects
— Part 1

Count Up to 10 Objects
— Part 2
How Many?
Practice

Chapter 7
Numbers to 10 — Part 2

6
7
8
9
10
0 to 10
Count and Match — Part 1
Count and Match — Part 2
Practice

PKB

Chapter 8
Ordinal Numbers

First
Second and Third
Fourth and Fifth
Practice

Chapter 9
Shapes and Solids

Cubes, Cylinders, and Spheres
Cubes
Positions
Build with Solids
Rectangles and Circles
Squares
Triangles

Squares, Circles,
Rectangles, and
Triangles — Part 1
Squares, Circles,
Rectangles, and
Triangles — Part 2
Practice

Chapter 10
Compare Sets

Match Objects
Which Set Has More?
Which Set Has Fewer?
More or Fewer?
Practice

Chapter 11
Compose and Decompose

Altogether — Part 1
Altogether — Part 2
Show Me
What's the Other Part? —
Part 1
What's the Other Part? —
Part 2
Practice

Chapter 12
Explore Addition and Subtraction

Add to 5 — Part 1
Add to 5 — Part 2
Two Parts Make a Whole
How Many in All?
Subtract Within 5 — Part 1
Subtract Within 5 — Part 2
How Many Are Left?

Practice

Chapter 13
Cumulative Review

Review 1 Match and Color
Review 2 Big and Small
Review 3 Heavy and Light
Review 4 Count to 5
Review 5 Count 5 Objects
Review 6 0
Review 7 Count Beads
Review 8 Patterns
Review 9 Length
Review 10 How Many?
Review 11 Ordinal Numbers
Review 12 Solids and
Shapes
Review 13 Which Set Has
More?
Review 14 Which Set Has
Fewer?
Review 15 Put Together
Review 16 Subtraction
Looking Ahead 1
Sequencing — Part 1
Looking Ahead 2
Sequencing — Part 2
Looking Ahead 3
Categorizing
Looking Ahead 4 Addition
Looking Ahead 5
Subtraction
Looking Ahead 6 Getting
Ready to Write Numerals
Looking Ahead 7 Reading
and Math

KA

Chapter 1
Match, Sort, and Classify

Left and Right
Same and Similar
Look for One That Is Different
How Does it Feel?
Match the Things That
Go Together
Sort
Practice

Chapter 2
Numbers to 5

Count to 5
Count Things Up to 5
Recognize the Numbers 1 to 3
Recognize the Numbers
4 and 5
Count and Match
Write the Numbers 1 and 2
Write the Number 3
Write the Number 4
Trace and Write 1 to 5
Zero
Picture Graphs
Practice

Chapter 3
Numbers to 10

Count 1 to 10
Count Up to 7 Things
Count Up to 9 Things
Count Up to 10 Things —
Part 1

Dimensions Math® Scope & Sequence

Count Up to 10 Things —
 Part 2
Recognize the Numbers
 6 to 10
Write the Numbers 6 and 7
Write the Numbers 8, 9,
 and 10
Write the Numbers 6 to 10
Count and Write the
 Numbers 1 to 10
Ordinal Positions
One More Than
Practice

Chapter 4
Shapes and Solids

Curved or Flat
Solid Shapes
Closed Shapes
Rectangles
Squares
Circles and Triangles
Where is It?
Hexagons
Sizes and Shapes
Combine Shapes
Graphs
Practice

Chapter 5
Compare Height, Length, Weight, and Capacity

Comparing Height
Comparing Length
Height and Length — Part 1
Height and Length — Part 2
Weight — Part 1

Weight — Part 2
Weight — Part 3
Capacity — Part 1
Capacity — Part 2
Practice

Chapter 6
Comparing Numbers Within 10

Same and More
More and Fewer
More and Less
Practice — Part 1
Practice — Part 2

KB

Chapter 7
Numbers to 20

Ten and Some More
Count Ten and Some More
Two Ways to Count
Numbers 16 to 20
Number Words 0 to 10
Number Words 11 to 15
Number Words 16 to 20
Number Order
1 More Than or Less Than
Practice — Part 1
Practice — Part 2

Chapter 8
Number Bonds

Putting Numbers Together
 — Part 1

Putting Numbers Together
 — Part 2
Parts Making a Whole
Look for a Part
Number Bonds for 2, 3, and 4
Number Bonds for 5
Number Bonds for 6
Number Bonds for 7
Number Bonds for 8
Number Bonds for 9
Number Bonds for 10
Practice — Part 1
Practice — Part 2
Practice — Part 3

Chapter 9
Addition

Introduction to Addition —
 Part 1
Introduction to Addition —
 Part 2
Introduction to Addition —
 Part 3
Addition
Count On — Part 1
Count On — Part 2
Add Up to 3 and 4
Add Up to 5 and 6
Add Up to 7 and 8
Add Up to 9 and 10
Addition Practice
Practice

Chapter 10
Subtraction

Take Away to Subtract —
 Part 1

Take Away to Subtract —
    Part 2
Take Away to Subtract —
    Part 3
Take Apart to Subtract —
    Part 1
Take Apart to Subtract —
    Part 2
Count Back
Subtract Within 5
Subtract Within 10 — Part 1
Subtract Within 10 — Part 2
Practice

Chapter 11
Addition and Subtraction

Add and Subtract
Practice Addition and
    Subtraction
Part-Whole Addition and
    Subtraction
Add to or Take Away
Put Together or Take Apart
Practice

Chapter 12
Numbers to 100

Count by Tens — Part 1
Count by Tens — Part 2
Numbers to 30
Numbers to 40
Numbers to 50
Numbers to 80
Numbers to 100 — Part 1
Numbers to 100 — Part 2
Count by Fives — Part 1
Count by Fives — Part 2

Practice

Chapter 13
Time

Day and Night
Learning About the Clock
Telling Time to the Hour —
    Part 1
Telling Time to the Hour —
    Part 2
Practice

Chapter 14
Money

Coins
Pennies
Nickels
Dimes
Quarters
Practice

1A

Chapter 1
Numbers to 10

Numbers to 10
The Number 0
Order Numbers
Compare Numbers
Practice

Chapter 2
Number Bonds

Make 6
Make 7
Make 8

Make 9
Make 10 — Part 1
Make 10 — Part 2
Practice

Chapter 3
Addition

Addition as Putting Together
Addition as Adding More
Addition with 0
Addition with Number Bonds
Addition by Counting On
Make Addition Stories
Addition Facts
Practice

Chapter 4
Subtraction

Subtraction as Taking Away
Subtraction as Taking Apart
Subtraction by Counting Back
Subtraction with 0
Make Subtraction Stories
Subtraction with Number
    Bonds
Addition and Subtraction
Make Addition and Subtraction
    Story Problems
Subtraction Facts
Practice
Review 1

Chapter 5
Numbers to 20

Numbers to 20
Add or Subtract Tens
    or Ones
Order Numbers to 20

Dimensions Math® Scope & Sequence

Compare Numbers to 20
Addition
Subtraction
Practice

Chapter 6
Addition to 20

Add by Making 10 — Part 1
Add by Making 10 — Part 2
Add by Making 10 — Part 3
Addition Facts to 20
Practice

Chapter 7
Subtraction Within 20

Subtract from 10 — Part 1
Subtract from 10 — Part 2
Subtract the Ones First
Word Problems
Subtraction Facts Within 20
Practice

Chapter 8
Shapes

Solid and Flat Shapes
Grouping Shapes
Making Shapes
Practice

Chapter 9
Ordinal Numbers

Naming Positions
Word Problems
Practice
Review 2

1B

Chapter 10
Length

Comparing Lengths Directly
Comparing Lengths Indirectly
Comparing Lengths with Units
Practice

Chapter 11
Comparing

Subtraction as Comparison
Making Comparison
  Subtraction Stories
Picture Graphs
Practice

Chapter 12
Numbers to 40

Numbers to 40
Tens and Ones
Counting by Tens and Ones
Comparing
Practice

Chapter 13
Addition and Subtraction Within 40

Add Ones
Subtract Ones
Make the Next Ten
Use Addition Facts
Subtract from Tens
Use Subtraction Facts
Add Three Numbers
Practice

Chapter 14
Grouping and Sharing

Adding Equal Groups
Sharing
Grouping
Practice

Chapter 15
Fractions

Halves
Fourths
Practice
Review 3

Chapter 16
Numbers to 100

Numbers to 100
Tens and Ones
Count by Ones or Tens
Compare Numbers to 100
Practice

Chapter 17
Addition and Subtraction Within 100

Add Ones — Part 1
Add Tens
Add Ones — Part 2
Add Tens and Ones — Part 1
Add Tens and Ones — Part 2
Subtract Ones — Part 1
Subtract from Tens
Subtract Ones — Part 2
Subtract Tens

Teacher's Guide KA

Subtract Tens and Ones —
    Part 1
Subtract Tens and Ones —
    Part 2
Practice

Chapter 18
Time

Telling Time to the Hour
Telling Time to the Half Hour
Telling Time to the 5 Minutes
Practice

Chapter 19
Money
Coins
Counting Money
Bills
Shopping
Practice
Review 4

2A

Chapter 1
Numbers to 1,000

Tens and Ones
Counting by Tens or Ones
Comparing Tens and Ones
Hundreds, Tens, and Ones
Place Value
Comparing Hundreds, Tens,
    and Ones
Counting by Hundreds, Tens,
    or Ones
Practice

Chapter 2
Addition and Subtraction — Part 1

Strategies for Addition
Strategies for Subtraction
Parts and Whole
Comparison
Practice

Chapter 3
Addition and Subtraction — Part 2

Addition Without Regrouping
Subtraction Without
    Regrouping
Addition with Regrouping
    Ones
Addition with Regrouping
    Tens
Addition with Regrouping
    Tens and Ones
Practice A
Subtraction with Regrouping
    from Tens
Subtraction with Regrouping
    from Hundreds
Subtraction with Regrouping
    from Two Places
Subtraction with Regrouping
    across Zeros
Practice B
Practice C

Chapter 4
Length

Centimeters
Estimating Length in
    Centimeters

Meters
Estimating Length in Meters
Inches
Using Rulers
Feet
Practice

Chapter 5
Weight

Grams
Kilograms
Pounds
Practice
Review 1

Chapter 6
Multiplication and Division

Multiplication — Part 1
Multiplication — Part 2
Practice A
Division — Part 1
Division — Part 2
Multiplication and Division
Practice B

Chapter 7
Multiplication and Division of 2, 5, and 10

The Multiplication Table of 5
Multiplication Facts of 5
Practice A
The Multiplication Table of 2
Multiplication Facts of 2
Practice B
The Multiplication Table of 10
Dividing by 2

Dimensions Math® Scope & Sequence

Dividing by 5 and 10
Practice C
Word Problems
Review 2

2B

Chapter 8
Mental Calculation

Adding Ones Mentally
Adding Tens Mentally
Making 100
Adding 97, 98, or 99
Practice A
Subtracting Ones Mentally
Subtracting Tens Mentally
Subtracting 97, 98, or 99
Practice B
Practice C

Chapter 9
Multiplication and Division of 3 and 4

The Multiplication Table of 3
Multiplication Facts of 3
Dividing by 3
Practice A
The Multiplication Table of 4
Multiplication Facts of 4
Dividing by 4
Practice B
Practice C

Chapter 10
Money

Making $1
Dollars and Cents
Making Change
Comparing Money
Practice A
Adding Money
Subtracting Money
Practice B

Chapter 11
Fractions

Halves and Fourths
Writing Unit Fractions
Writing Fractions
Fractions that Make 1 Whole
Comparing and Ordering
    Fractions
Practice
Review 3

Chapter 12
Time

Telling Time
Time Intervals
A.M. and P.M.
Practice

Chapter 13
Capacity

Comparing Capacity
Units of Capacity
Practice

Chapter 14
Graphs

Picture Graphs
Bar Graphs
Practice

Chapter 15
Shapes

Straight and Curved Sides
Polygons
Semicircles and Quarter-
    circles
Patterns
Solid Shapes
Practice
Review 4
Review 5

3A

Chapter 1
Numbers to 10,000

Numbers to 10,000
Place Value — Part 1
Place Value — Part 2
Comparing Numbers
The Number Line
Practice A
Number Patterns
Rounding to the Nearest
    Thousand
Rounding to the Nearest
    Hundred
Rounding to the Nearest Ten
Practice B

Chapter 2
Addition and Subtraction — Part 1

Mental Addition — Part 1
Mental Addition — Part 2
Mental Subtraction — Part 1
Mental Subtraction — Part 2
Making 100 and 1,000
Strategies for Numbers Close
    to Hundreds
Practice A
Sum and Difference
Word Problems — Part 1
Word Problems — Part 2
2-Step Word Problems
Practice B

Chapter 3
Addition and Subtraction — Part 2

Addition with Regrouping
Subtraction with Regrouping
    — Part 1
Subtraction with Regrouping
    — Part 2
Estimating Sums and
    Differences — Part 1
Estimating Sums and
    Differences — Part 2
Word Problems
Practice

Chapter 4
Multiplication and Division

Looking Back at
    Multiplication
Strategies for Finding the
    Product
Looking Back at Division
Multiplying and Dividing with
    0 and 1
Division with Remainders
Odd and Even Numbers
Word Problems — Part 1
Word Problems — Part 2
2-Step Word Problems
Practice
Review 1

Chapter 5
Multiplication

Multiplying Ones, Tens, and
    Hundreds
Multiplication Without
    Regrouping
Multiplication with
    Regrouping Tens
Multiplication with
    Regrouping Ones
Multiplication with
    Regrouping Ones and
    Tens
Practice A
Multiplying a 3-Digit Number
    with Regrouping Once
Multiplication with Regrouping
    More Than Once
Practice B

Chapter 6
Division

Dividing Tens and Hundreds
Dividing a 2-Digit Number
    by 2 — Part 1
Dividing a 2-Digit Number
    by 2 — Part 2
Dividing a 2-Digit Number by
    3, 4, and 5
Practice A
Dividing a 3-Digit Number by 2
Dividing a 3-Digit Number by
    3, 4, and 5
Dividing a 3-Digit Number,
    Quotient is 2 Digits
Practice B

Chapter 7
Graphs and Tables

Picture Graphs and Bar Graphs
Bar Graphs and Tables
Practice
Review 2

3B

Chapter 8
Multiplying and Dividing with 6, 7, 8, and 9

The Multiplication Table of 6
The Multiplication Table of 7
Multiplying by 6 and 7
Dividing by 6 and 7
Practice A
The Multiplication Table of 8

Dimensions Math® Scope & Sequence

The Multiplication Table of 9
Multiplying by 8 and 9
Dividing by 8 and 9
Practice B

Chapter 9
Fractions — Part 1

Fractions of a Whole
Fractions on a Number Line
Comparing Fractions with
    Like Denominators
Comparing Fractions with
    Like Numerators
Practice

Chapter 10
Fractions — Part 2

Equivalent Fractions
Finding Equivalent Fractions
Simplifying Fractions
Comparing Fractions — Part 1
Comparing Fractions — Part 2
Practice A
Adding and Subtracting
    Fractions — Part 1
Adding and Subtracting
    Fractions — Part 2
Practice B

Chapter 11
Measurement

Meters and Centimeters
Subtracting from Meters
Kilometers
Subtracting from Kilometers
Liters and Milliliters
Kilograms and Grams

Word Problems
Practice
Review 3

Chapter 12
Geometry

Circles
Angles
Right Angles
Triangles
Properties of Triangles
Properties of Quadrilaterals
Using a Compass
Practice

Chapter 13
Area and Perimeter

Area
Units of Area
Area of Rectangles
Area of Composite Figures
Practice A
Perimeter
Perimeter of Rectangles
Area and Perimeter
Practice B

Chapter 14
Time

Units of Time
Calculating Time — Part 1
Practice A
Calculating Time — Part 2
Calculating Time — Part 3
Calculating Time — Part 4
Practice B

Chapter 15
Money

Dollars and Cents
Making $10
Adding Money
Subtracting Money
Word Problems
Practice
Review 4
Review 5

4A

Chapter 1
Numbers to One Million

Numbers to 100,000
Numbers to 1,000,000
Number Patterns
Comparing and Ordering
    Numbers
Rounding 5-Digit Numbers
Rounding 6-Digit Numbers
Calculations and Place Value
Practice

Chapter 2
Addition and Subtraction

Addition
Subtraction
Other Ways to Add and
    Subtract — Part 1
Other Ways to Add and
    Subtract — Part 2
Word Problems

Practice

Chapter 3
Multiples and Factors

Multiples
Common Multiples
Factors
Prime Numbers and
    Composite Numbers
Common Factors
Practice

Chapter 4
Multiplication

Mental Math for Multiplication
Multiplying by a 1-Digit
    Number — Part 1
Multiplying by a 1-Digit
    Number — Part 2
Practice A
Multiplying by a Multiple of 10
Multiplying by a 2-Digit
    Number — Part 1
Multiplying by a 2-Digit
    Number — Part 2
Practice B

Chapter 5
Division

Mental Math for Division
Estimation and Division
Dividing 4-Digit Numbers
Practice A
Word Problems
Challenging Word Problems
Practice B
Review 1

Chapter 6
Fractions

Equivalent Fractions
Comparing and Ordering
    Fractions
Improper Fractions and Mixed
    Numbers
Practice A
Expressing an Improper
    Fraction as a Mixed
    Number
Expressing a Mixed Number
    as an Improper Fraction
Fractions and Division
Practice B

Chapter 7
Adding and Subtracting Fractions

Adding and Subtracting
    Fractions — Part 1
Adding and Subtracting
    Fractions — Part 2
Adding a Mixed Number and
    a Fraction
Adding Mixed Numbers
Subtracting a Fraction from
    a Mixed Number
Subtracting Mixed Numbers
Practice

Chapter 8
Multiplying a Fraction and a Whole Number

Multiplying a Unit Fraction
    by a Whole Number

Multiplying a Fraction by a
    Whole Number — Part 1
Multiplying a Fraction by a
    Whole Number — Part 2
Fraction of a Set
Multiplying a Whole Number
    by a Fraction — Part 1
Multiplying a Whole Number
    by a Fraction — Part 2
Word Problems — Part 1
Word Problems — Part 2
Practice

Chapter 9
Line Graphs and Line Plots

Line Graphs
Drawing Line Graphs
Line Plots
Practice
Review 2

4B

Chapter 10
Measurement

Metric Units of Measurement
Customary Units of Length
Customary Units of Weight
Customary Units of Capacity
Units of Time
Practice A
Fractions and Measurement
    — Part 1
Fractions and Measurement
    — Part 2
Practice B

Dimensions Math® Scope & Sequence

Chapter 11
Area and Perimeter

Area of Rectangles — Part 1
Area of Rectangles — Part 2
Area of Composite Figures
Perimeter — Part 1
Perimeter — Part 2
Practice

Chapter 12
Decimals

Tenths — Part 1
Tenths — Part 2
Hundredths — Part 1
Hundredths — Part 2
Expressing Decimals as
    Fractions in Simplest Form
Expressing Fractions as
    Decimals
Practice A
Comparing and Ordering
    Decimals
Rounding Decimals
Practice B

Chapter 13
Addition and Subtraction of Decimals

Adding and Subtracting Tenths
Adding Tenths with Regrouping
Subtracting Tenths with
    Regrouping
Practice A
Adding Hundredths
Subtracting from 1 and 0.1
Subtracting Hundredths
Money, Decimals, and Fractions

Practice B
Review 3

Chapter 14
Multiplication and Division of Decimals

Multiplying Tenths and
    Hundredths
Multiplying Decimals by a
    Whole Number — Part 1
Multiplying Decimals by a
    Whole Number — Part 2
Practice A
Dividing Tenths and Hundredths
Dividing Decimals by a Whole
    Number — Part 1
Dividing Decimals by a Whole
    Number — Part 2
Dividing Decimals by a Whole
    Number — Part 3
Practice B

Chapter 15
Angles

The Size of Angles
Measuring Angles
Drawing Angles
Adding and Subtracting Angles
Reflex Angles
Practice

Chapter 16
Lines and Shapes

Perpendicular Lines
Parallel Lines
Drawing Perpendicular and
    Parallel Lines
Quadrilaterals

Lines of Symmetry
Symmetrical Figures and
    Patterns
Practice

Chapter 17
Properties of Cuboids

Cuboids
Nets of Cuboids
Faces and Edges of Cuboids
Practice
Review 4
Review 5

5A

Chapter 1
Whole Numbers

Numbers to One Billion
Multiplying by 10, 100, and
    1,000
Dividing by 10, 100, and 1,000
Multiplying by Tens,
    Hundreds, and Thousands
Dividing by Tens, Hundreds,
    and Thousands
Practice

Chapter 2
Writing and Evaluating Expressions

Expressions with Parentheses
Order of Operations — Part 1
Order of Operations — Part 2

Other Ways to Write and
    Evaluate Expressions
Word Problems — Part 1
Word Problems — Part 2
Practice

Chapter 3
Multiplication and Division

Multiplying by a 2-digit
    Number — Part 1
Multiplying by a 2-digit
    Number — Part 2
Practice A
Dividing by a Multiple of Ten
Divide a 2-digit Number by a
    2-digit Number
Divide a 3-digit Number by a
    2-digit Number — Part 1
Divide a 3-digit Number by a
    2-digit Number — Part 2
Divide a 4-digit Number by a
    2-digit Number
Practice B

Chapter 4
Addition and Subtraction of Fractions

Fractions and Division
Adding Unlike Fractions
Subtracting Unlike Fractions
Practice A
Adding Mixed Numbers
    — Part 1
Adding Mixed Numbers
    — Part 2
Subtracting Mixed Numbers
    — Part 1

Subtracting Mixed Numbers
    — Part 2
Practice B
Review 1

Chapter 5
Multiplication of Fractions

Multiplying a Fraction by a
    Whole Number
Multiplying a Whole Number
    by a Fraction
Word Problems — Part 1
Practice A
Multiplying a Fraction by a
    Unit Fraction
Multiplying a Fraction by a
    Fraction — Part 1
Multiplying a Fraction by a
    Fraction — Part 2
Multiplying Mixed Numbers
Word Problems — Part 2
Fractions and Reciprocals
Practice B

Chapter 6
Division of Fractions

Dividing a Unit Fraction by a
    Whole Number
Dividing a Fraction by a
    Whole Number
Practice A
Dividing a Whole Number by
    a Unit Fraction
Dividing a Whole Number by
    a Fraction
Word Problems
Practice B

Chapter 7
Measurement

Fractions and Measurement
    Conversions
Fractions and Area
Practice A
Area of a Triangle — Part 1
Area of a Triangle — Part 2
Area of Complex Figures
Practice B

Chapter 8
Volume of Solid Figures

Cubic Units
Volume of Cuboids
Finding the Length of an Edge
Practice A
Volume of Complex Shapes
Volume and Capacity — Part 1
Volume and Capacity — Part 2
Practice B
Review 2

5B

Chapter 9
Decimals

Thousandths
Place Value to Thousandths
Comparing Decimals
Rounding Decimals
Practice A
Multiply Decimals by 10, 100,
    and 1,000
Divide Decimals by 10, 100,
    and 1,000

Dimensions Math® Scope & Sequence

Conversion of Measures
Mental Calculation
Practice B

Chapter 10
The Four Operations of Decimals

Adding Decimals to
     Thousandths
Subtracting Decimals
Multiplying by 0.1 or 0.01
Multiplying by a Decimal
Practice A
Dividing by a Whole Number
     — Part 1
Dividing by a Whole Number
     — Part 2
Dividing a Whole Number by
     0.1 and 0.01
Dividing a Whole Number by
     a Decimal
Practice B

Chapter 11
Geometry

Measuring Angles
Angles and Lines
Classifying Triangles
The Sum of the Angles in a
     Triangle
The Exterior Angle of a
     Triangle
Classifying Quadrilaterals
Angles of Quadrilaterals
     — Part 1
Angles of Quadrilaterals
     — Part 2

Drawing Triangles and
     Quadrilaterals
Practice

Chapter 12
Data Analysis and Graphs

Average — Part 1
Average — Part 2
Line Plots
Coordinate Graphs
Straight Line Graphs
Practice
Review 3

Chapter 13
Ratio

Finding the Ratio
Equivalent Ratios
Finding a Quantity
Comparing Three Quantities
Word Problems
Practice

Chapter 14
Rate

Finding the Rate
Rate Problems — Part 1
Rate Problems — Part 2
Word Problems
Practice

Chapter 15
Percentage

Meaning of Percentage
Expressing Percentages
     as Fractions

Percentages and Decimals
Expressing Fractions as
     Percentages
Practice A
Percentage of a Quantity
Word Problems
Practice B
Review 4
Review 5

Teacher's Guide KA

Suggested number of class periods: 7 – 8

Lesson		Page	Resources		Objectives
	Chapter Opener	p. 5	TB:	p. 1	
1	Left and Right	p. 6	TB: WB:	p. 2 p. 1	Distinguish left and right.
2	Same and Similar	p. 8	TB: WB:	p. 4 p. 3	Identify similar objects and say in what way they are the same.
3	Look for One That is Different	p. 10	TB: WB:	p. 6 p. 5	Identify objects that are different.
4	How Does it Feel?	p. 12	TB: WB:	p. 8 p. 7	Use texture words to describe similarities and differences in objects.
5	Match the Things That Go Together	p. 14	TB: WB:	p. 10 p. 9	Identify objects that go together.
6	Sort	p. 16	TB: WB:	p. 12 p. 11	Sort objects in more than one way.
7	Practice	p. 17	TB: WB:	p. 13 p. 13	Practice concepts from the chapter.
	Workbook Solutions	p. 19			

This chapter builds on the content in Chapter 1 of the same name in **Dimensions Math® Pre-Kindergarten A** and **B**. Students in Kindergarten will briefly review matching and sorting.

The emphasis in this chapter is on having students explain their reasoning. Students will classify, sort, and compare objects by:

- Color
- Texture or feel
- Shape
- Size
- Orientation or rotation

Matching, sorting, and classifying are important prerequisite skills for counting. In the next chapter, students will be counting in pictures and will first need to sort in their minds. On page 17, students count shovels, but not pails. They count all white objects or objects that are animals or circles.

Students may have different ideas of how to match, sort, and classify objects. They should be able to explain their reasoning. Encourage students to use math vocabulary in their explanations. These skills will be further developed in **Chapter 4: Shapes** and **Chapter 5: Compare Height, Length, Weight, and Capacity**.

You may be able to move through this chapter quickly, depending on your students' background knowledge. Two lessons may be covered in one math period.

Storybooks that go along with content in this chapter can be added as time allows or for additional exploration. Storybook suggestions are listed on the following page.

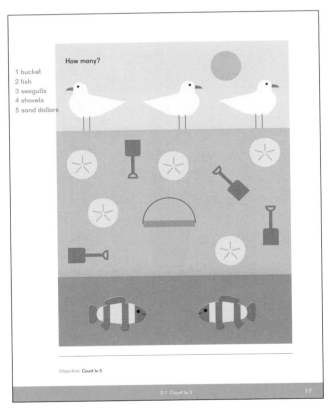

Dimensions Math® KA
Chapter 2, Lesson 1: Count to 5

Teacher's Guide KA Chapter 1

© 2017 Singapore Math Inc.

Materials

- Attribute bears
- Buttons
- Objects with rough and smooth textures
- Dice

Suggestions For Lesson 5: Match Things that Go Together

- Paper plates, cups, and napkins
- Pot and pot lid
- Plastic fork, knife, spoon
- Toy tools (hammer, screwdriver, etc.)
- Comb, hair ribbon, shampoo, hairbrush
- Small carton of milk and a paper cup
- Baby food jar, small spoon, baby bottle, pacifier
- Dog collar, leash, dog toy, dog treat
- Bracelet, necklace, earrings
- Greeting card, envelope, postage stamp, pen
- Box of cereal and cereal bowl
- Lunch box, thermos, small plastic bag holding carrots or apple slices
- Marker, crayon, art paper
- Coat, gloves, winter hat
- Sunglasses, beach towel, bathing suit
- Can of soup and package of crackers

Blackline Masters

- Matching Picture Cards
- Which One Doesn't Belong?
- Sudoku

Storybooks

- *Frog and Toad are Friends* by Arnold Lobel
- *The Button Box* by Margarette S. Reid
- *Wacky Wednesday* by Dr. Seuss
- *Nothing Like a Puffin* by Sue Soltis
- *Two Eggs, Please* by Sarah Weeks

Technology

Classrooms that have tablet devices available can use the camera application to record or document different activities at centers.

Letters Home

- Chapter 1 Letter

Notes

Explore

Show students items in the classroom that are identical (exactly the same), alike, or different. Include items that go together, like the mittens on the textbook page.

Have students share their ideas and ask:

- Why do the items go together?
- How are the items alike and different?

Examples:

- Pencil and paper are alike because they are both for writing. They are different because you write with one and write on the other.
- Pencils and pens are alike because they're both used to write. They are different because one is pencil lead and the other is ink.
- Unsharpened pencils are the same.
- Sharpened pencils are alike because they're both pencils but one is longer than the other.

Take students on a scavenger hunt either in the classroom or through the school and playground looking for things that go together and tell how they are alike and different.

Learn

Have students look at page 1. Ask them to identify the objects on the page. Ask, "What is alike and what is different about the objects?"

Chapter 1

Match, Sort, and Classify

1

Lesson 1 Left and Right

Objective

- Distinguish left and right.

Explore

Have students line up shoulder to shoulder. Tell them that you are going to give them a series of directions, and when you are done, they should all be facing the same way and near each other:

1. Take two steps.
2. Turn.
3. Take one step.
4. Turn.
5. Take three steps.
6. Turn.

Students should notice that without being told which direction to turn, they will not be facing the same way.

Have students line up single file, facing the board. Standing in the front of the line, also facing the board, raise your left hand and make an L shape. Ask students to do the same. Have them make the letter L, for "left," with their left hand.

While students are still standing in a single file line, point out the left and right sides of the room. Have students name objects which are on the left side of the room. Have them each raise their left hand, then their left foot, then point to their left ear. Finally, have students turn to the left.

Remind students which is the right side of the room from where they are facing. Then have them repeat the above activity but omit making the letter L shape with the right hand.

Play the turn game again with more explicit instructions:

- Take two steps to the left.
- Take one step backwards.
- Turn to the right.
- Take three steps forward.
- Turn to the left.
- Take one step to the right.

Lesson 1
Left and Right ①

Look and talk.
Which one is on the left?
Which one is on the right?

Objective: Distinguish left and right.

2   1-1 Left and Right

Learn

Show students two objects next to each other. Have them point to the one on the left or right.

On textbook page 2, have students identify items that are on the right and left:

- The bench is on the right side of the page.
- The blue slide is to the left of the orange slide.

Whole Group Activity

▲ **Hokey Pokey**

Teach students the Hokey Pokey (VR).

Small Group Activities

▲ **Textbook** Page 3

When students pick up their crayon to color the objects, identify if students are right or left-handed.

▲ **Copy Me Clap-Stomp**

Player 1 taps a pattern of 3 moves using either his hands or his feet. Player 2 stands next to Player 1 and copies the pattern. For example their pattern may be: Left foot stomp, right hand tap, left foot stomp.

▲ **Copy Me Blocks**

Materials: Pattern blocks

Player 1 uses up to 3 pattern blocks to create a design, keeping it hidden from Player 2. Player 1 then describes the design using words and phrases such as left and right, on top of, below, etc. and Player 2 tries to duplicate the design.

▲ **Left Hand Bracelet**

Materials: Yarn or string, beads

Have students string some beads on yarn or string to make a "left-hand" bracelet to wear for the week as a reminder of which hand is left and which is right.

Exercise 1 · page 1

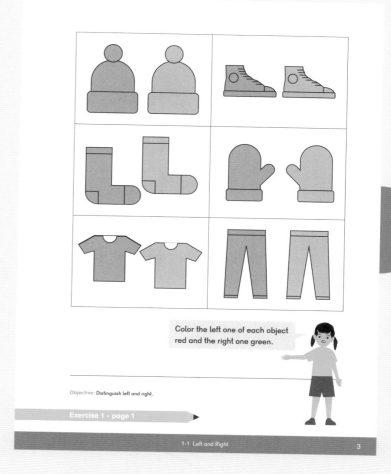

Color the left one of each object red and the right one green.

Objective: Distinguish left and right.

Exercise 1 · page 1

1-1 Left and Right   3

Extend

★ **Robot Map**

Player 1 plays the role of Programmer and Player 2 takes the role of Robot. The Programmer makes a list of steps that the Robot will need to make in order to get to a designated point.

L = Turn left	F = 1 step forward
R = Turn right	B = 1 step backwards

After the Programmer has mapped out the route, she reads the direction to the Robot, who then moves as commanded until she gets to the finishing point. Partners switch roles and play again.

Lesson 2 Same and Similar

Objective

- Identify similar objects and say in what way they are the same.

Lesson Materials

- Attribute bears
- Stuffed bear

Explore

Show students a stuffed bear and an attribute bear. Tell students that these are not exactly the same. Give examples of why the bears are not exactly the same. For example, one bear is bigger than the other, one is softer than the other, etc. Have students suggest other ways the bears are not the same.

Tell students the bears are the same in some ways. For example, both are bears, so they are the same that way. Ask students what else is the same about the bears.

Give pairs of students a handful of attribute bears. Ask the partners to find and show two bears that are the same color.

Next, have students find two bears that are similar, but not exactly the same. Define the word "similar," and have students explain how the two bears are similar. Examples:

- They are both red, but one is bigger than the other.
- They are both the same size, but one is red and the other is yellow.

Look and talk.
Which ones are the same?
Which ones are similar?

Objective: Find pairs of objects that are the same and that are similar.

4
1-2 Same and Similar

Learn

Have students look at textbook page 4. Ask them to name the objects on the page. Have them tell how the objects are the same or similar:

- The backpacks are similar, but different colors.
- The backpacks are similar, but different sizes.
- Two of the coats are the same, but the third one has no star on it.

Teacher's Guide KA Chapter 1

Whole Group Activity

▲ I Spy — Same and Different

With students in a circle, say, "I spy with my little eye two things that are exactly the same." Then give clues regarding the objects.

The student who guesses the objects correctly gives the next clues. Repeat for objects that are not exactly the same.

Small Group Activities

▲ Textbook Page 5

Have students draw a line with their finger first, then a pencil or crayon.

▲ Match

Materials: Matching Picture Cards (BLM)

Students arrange the cards, faceup, in a grid. Students take turns finding two cards that go together.

★ Memory

Materials: Matching Picture Cards (BLM)

Students arrange the cards, facedown, in a grid. Students take turns finding two cards that go together.

Exercise 2 • page 3

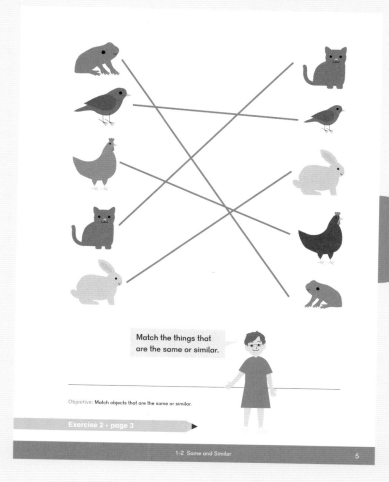

Match the things that are the same or similar.

Objective: Match objects that are the same or similar.

Exercise 2 • page 3

1-2 Same and Similar   5

Extend

★ Similar Game Cards

Materials: Paper, cut into quarters

Provide students with $\frac{1}{4}$-page size sheets of paper. Have them create pairs of cards that have an object on each that is either:

*   Exactly the same
*   Similar, but a different size, color or direction (see textbook page 5 for examples)

Laminate the cards and use for **Match** or **Memory** games in the classroom.

Objective

- Identify objects that are different.

Lesson Materials

- Classroom objects for comparison

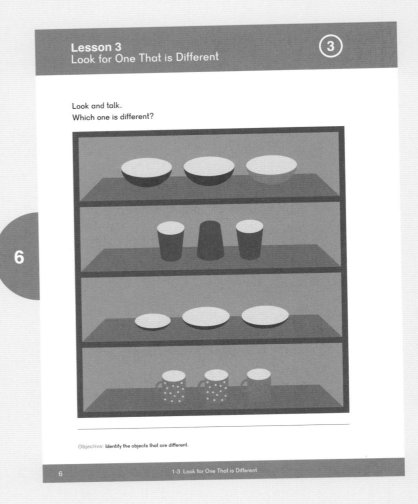

Explore

Have students find a partner in the classroom who has something the same or similar about them. Partners share their reasons for what they have that is the same.

Examples:

- Same or similar color of hair or shirt
- Both wear glasses
- Both are girls
- They are the same height
- They're wearing the same type of shoes

Now have partners share something about each other that is different.

Examples:

- My pants are long and my partner's are short.
- My t-shirt is red and my partner's is blue.

Learn

Have the students look at page 6 and identify the objects on the page. Ask them to discuss the objects in each row. Ask, "Which one is different? How is it different?"

Whole Group Activity

▲ Same Different Dance

A Leader demonstrates dance moves and tells students to do either the same or different steps.

For example, the Leader marches and says, "Same," so all students march in place. Then the Leader changes to twirls and says, "Different," so students must do a different move from the leader.

Small Group Activities

▲ Textbook Page 7

Have students circle the picture that is different with their finger first, then a pencil.

▲ What's Different?

Materials: Classroom objects

Player 1 sets out objects. Player 2 reviews them, then closes her eyes or turns her back. Player 1 makes a change to one of the objects. Player 2 looks again and says how the objects are now different.

Exercise 3 • page 5

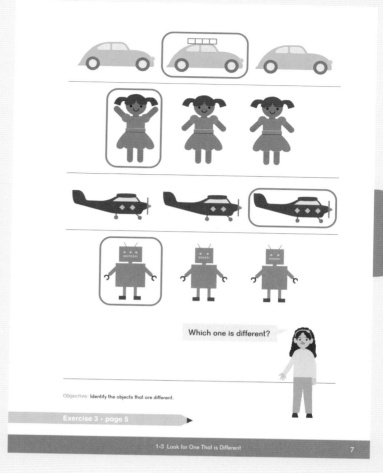

Which one is different?

Objective: Identify the objects that are different.

Exercise 3 • page 5

1-3 Look for One That is Different

7

Extend

★ Different Pictures

Materials: Art supplies such as dot markers or stickers

Have students create their own "find things that are different" pictures. Fold a sheet of paper in half and draw, use dot markers, or stickers to create a simple scene on one half. On the other half, create the same scene with a few changes (different color flowers, one more fish, etc.).

Students can trade with classmates and have them find the differences, or their pictures can be collected for a class book.

Lesson 4 How Does it Feel?

Objective

- Use texture words to describe similarities and differences in objects.

Lesson Materials

- Brown paper bags with a smooth or rough object in each — 1 bag per pair of students (a long sock would also work instead of the paper bag)

Explore

Pass out bags containing a small object to each pair of students. Without looking, have one student describe what is in the bag to his partner.

The student that described takes the object out of the bag, and students discuss the description and whether his guess was correct or incorrect.

Have partners trade bags with another pair. This time the person who described is now the guesser.

Discuss the different ways that objects can be described. Students have worked with size, color, and position. Tell them that the way things feel is another way to describe something. This is called the "texture." Have students share the different texture words they used to describe the objects in their bags.

Learn

Have the students look at page 8 and discuss which items on the table are rough and which are smooth.

Ask if students know other words to describe the textures of the items:

- Pineapples can be bumpy or prickly.
- Bread is crusty on the outside and soft on the inside.

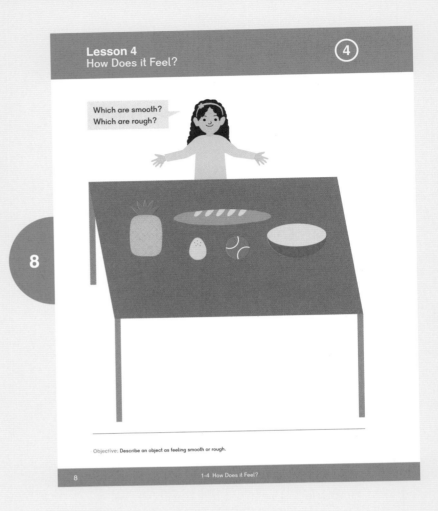

Whole Group Activity

▲ Guessing Game

The Leader thinks of an object. Students take turns asking a question about the object that can be answered with "Yes" or "No," until the object is identified.

For example, the Leader is thinking of a kitten. The first student might ask, "Is what you are thinking of soft?" The Leader responds, "Yes."

Small Group Activities

▲ **Textbook** Page 9

▲ **Match It**

Materials: Ribbons, paper, rocks, interlocking bricks, bag or sock

Use pairs of objects that have different textures such as ribbons, paper, rocks, interlocking bricks, etc. Place one set of the objects in a bag or sock and leave the other set out to see.

Player 1 reaches into the bag, feels one of the objects, and describes it to Player 2. Player 2 tries to match the object being described by Player 1 to the object that can be seen.

Take it Outside

▲ **Handmade Jewels (Rough and Smooth)**

Materials: Rock salt, food coloring, white glue

Prepare a mixture of 2 cups of rock salt and 6 – 8 drops of food coloring together. Add $\frac{1}{2}$ cup of white glue and continue mixing. Students mix and sculpt into jewels. Allow to dry.

▲ **Kinetic Sand (Smooth)**

Materials: Flour, baby oil, gallon storage bags

Combine 8 cups of flour and 1 cup of baby oil in gallon zipper bags, seal, and mix. Students can mold and sculpt.

Exercise 4 · page 7 ▶

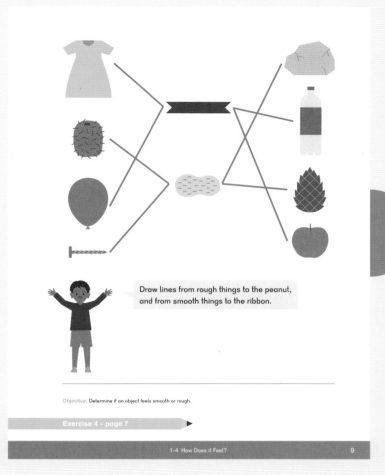

Draw lines from rough things to the peanut, and from smooth things to the ribbon.

Objective: Determine if an object feels smooth or rough.

Exercise 4 · page 7 ▶

1-4 How Does it Feel?   9

Extend

★ **Smooth and Rough Ramps**

Materials: Blocks, sturdy pieces of cardboard or large books, materials with different textures such as sandpaper, cloth, or carpet, toy cars, balls, and/or cylinders

Have students build a ramp with blocks and a sturdy piece of cardboard, then cover the ramp with sandpaper, cloth, carpet, etc.

Have students roll a toy car or round object down the ramps and notice how the texture affects the speed of the object as it rolls.

Lesson 5 Match the Things That Go Together

Objective

- Identify objects that go together.

Lesson Materials

- Things that go together (see **Chapter Materials** on page 3 of this Teacher's Guide for suggestions)

Explore

Set out the materials that go together on the floor in the middle of a circle of students.

Call on two or three students at a time to find two objects that go together. Discuss students' objects. Ask students:

- Why do the objects go together?
- How are they different?

Repeat until all students have had a turn.

Have students discuss if there are other objects still on the floor that could go with the pairs they have already made. Have them explain their reasoning, then add the object to the set.

Discuss with students if any of the objects could be moved from one group to another. Have them explain why the objects belong together.

Learn

Have students look at page 10 and identify the objects on the page. Ask students, "Which things go together?" Have them explain why they think the objects go together.

While the obvious answer may seem to be, "The frog and lily pad go together," students may say that the frog, duck, and duckling go together because they all go in the water.

Whole Group Activity

▲ **Match Up**

Materials: Matching Picture Cards (BLM)

Pass out a set of Matching Picture Cards (BLM) with objects that go together, one card for each student. On your signal, students organize themselves into groups or pairs of students that have pictures of objects that go together. Have groups explain their thinking.

Small Group Activities

▲ **Textbook** Page 11

▲ **Match**

Materials: Matching Picture Cards (BLM)

Students arrange the cards, faceup, in a grid.
Students take turns finding two cards that go together.

★ **Memory**

Materials: Matching Picture Cards (BLM)

Students arrange the cards, facedown, in a grid.
Students take turns finding two cards that go together.

Exercise 5 • page 9

Extend

★ **Which One Doesn't Belong?**

The picture shown to the right is also used on the
Which One Doesn't Belong? (BLM) worksheet.

Find a reason why each one does not belong.
Possible student answers:

· One dog is facing left and the rest are facing right.
· One dog is white and the rest are black.
· One dog is smaller than the others.
· One dog has spots and the rest don't.

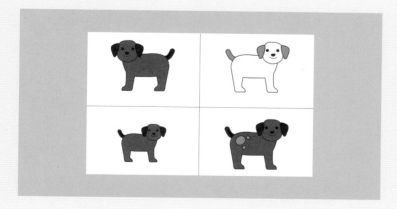

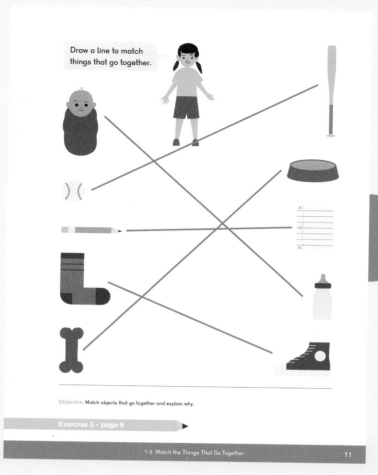

Draw a line to match things that go together.

Objective: Match objects that go together and explain why.

Exercise 5 • page 9

1-5 Match the Things That Go Together   11

Lesson 6 Sort

Objective

- Sort objects in more than one way.

Lesson Materials

- Buttons to be sorted in zipper bags (ask parents to donate buttons, if necessary)

Explore

Tell students that so far in this chapter, they have grouped objects in different ways: size, color, texture and things that go together.

Give bags of buttons to pairs of students. Have them sort them into groups. Have students share some of their reasons for sorting the buttons. Examples:

- 2 holes and 4 holes
- Size, shape, texture, or color
- Shank buttons and buttons with holes

After students have shared, ask them to sort the same pile of buttons a different way.

Learn

Have the students look at page 12. For each row, ask students, "Which one is different?"

Whole Group Activity

▲ **What Was I Thinking?**

Sort students into groups, and have them identify the reason for the sort. Some examples for sorting could include hair color, shirt color, and glasses vs. no glasses.

Exercise 8 • page 11

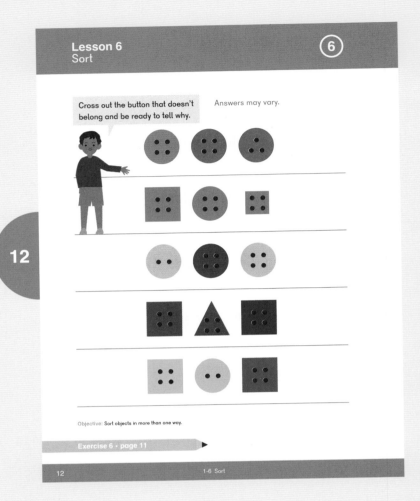

Objective

- Practice concepts from the chapter.

Practice days are designed for further practice and assessment as needed.

Students can complete the textbook pages and workbook pages as practice, as assessment or choose one or the other.

Use **Activities** from the chapter for students not being assessed.

Small Group Activity

▲ **Left-Right-Center**

Materials: Modified die, counters

Modify a die to have sides of L, R, C, and three sides showing ●.

Players sit in a circle and begin with two counters each. They take turns rolling the die. On each roll, the player either keeps or passes a counter, depending on the roll:

- L: Pass one of the counters to the left
- R: Pass one of the counters to the right
- C: Put one counter in the center
- ●: Keep both counters

Play moves to the player on the left after each roll.

Players who have no counters left are still in the game as players on either side can still pass counters. They just do not roll the dice until they have counter(s) again. When all but one of the counters are in the center, the player still holding that remaining counter is the winner.

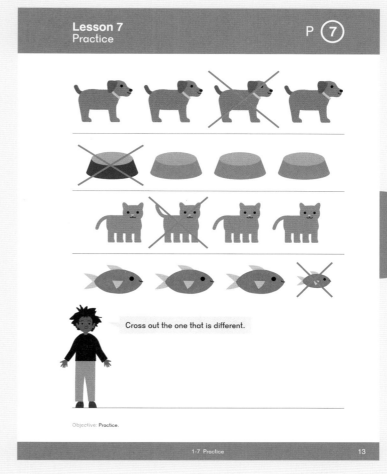

Lesson 7
Practice

P ⑦

Cross out the one that is different.

Objective: Practice.

1-7 Practice

13

Extend

★ Button Sudoku

Materials: Sudoku (BLM)

Give each student a copy of Sudoku (BLM). To play, cut out the pictures and put one in the correct square.

Each button appears only once in each:

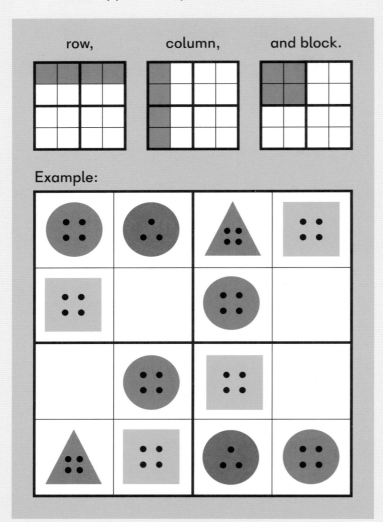

row,   column,   and block.

Example:

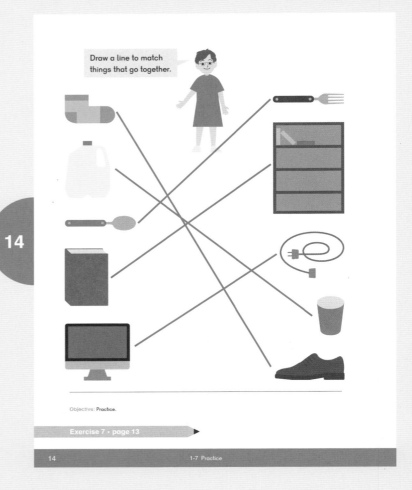

Draw a line to match things that go together.

Objective: **Practice.**

Exercise 7 • page 13

14   1-7 Practice

14

Chapter 1 Match, Sort, and Classify

Exercise 1

Circle the animals facing left.

Using this page: Have students identify the animals that are facing the left side of the page and circle.
Concept: Identifying left.

1-1 Left and Right
1

Color the things on the left.

Color the things on the right.

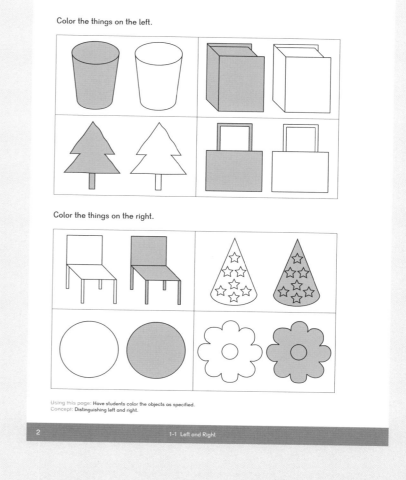

Using this page: Have students color the objects as specified.
Concept: Distinguishing left and right.

2
1-1 Left and Right

Exercise 2

Circle the 2 penguins that are the same.

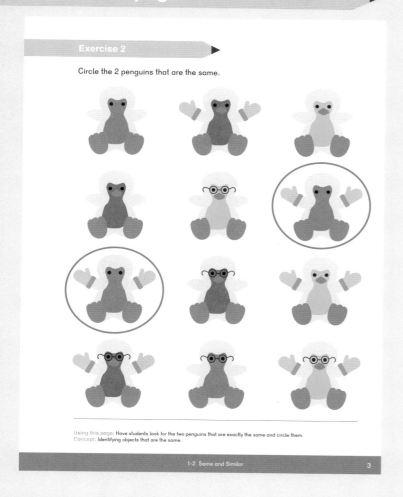

Using this page: Have students look for the two penguins that are exactly the same and circle them.
Concept: Identifying objects that are the same.

1-2 Same and Similar
3

Circle the things that are similar.

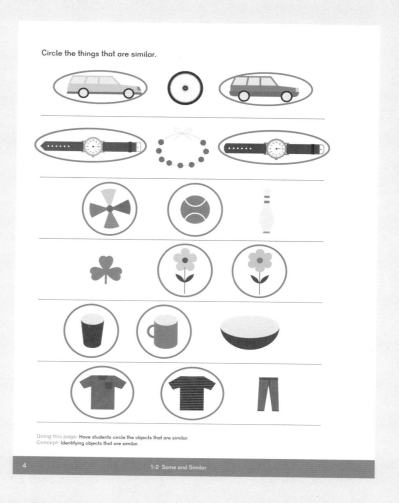

Using this page: Have students circle the objects that are similar.
Concept: Identifying objects that are similar.

4
1-2 Same and Similar

Exercise 3 • pages 5 – 6

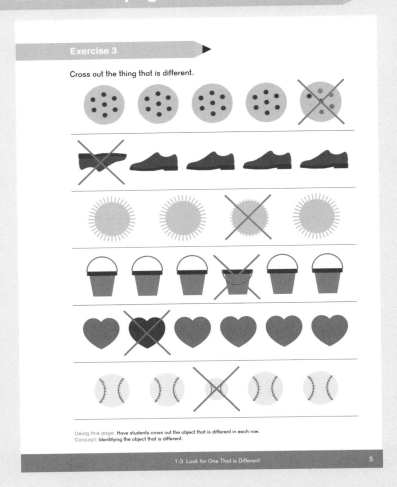

Exercise 4 • pages 7 – 8

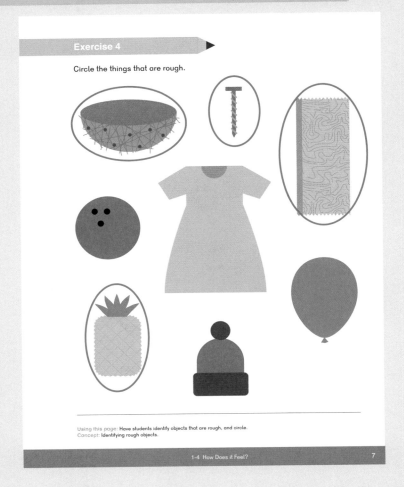

Exercise 5

Match.

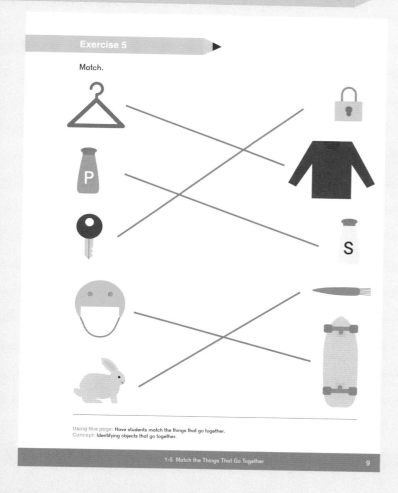

Using this page: Have students match the things that go together.
Concept: Identifying objects that go together.

1-5 Match the Things That Go Together   9

Circle the things that go together.

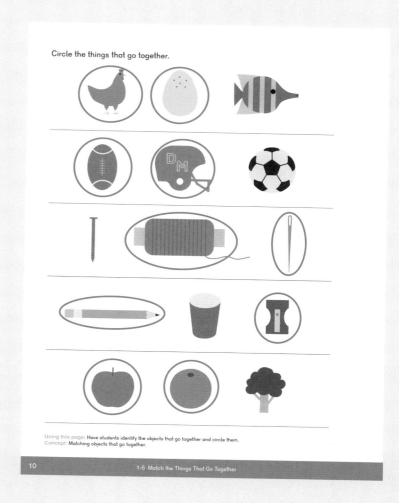

Using this page: Have students identify the objects that go together and circle them.
Concept: Matching objects that go together.

10   1-5 Match the Things That Go Together

Exercise 6

Sort the buttons. This answer is based on sorting by color. Students' answer may differ since there is more than one way to sort

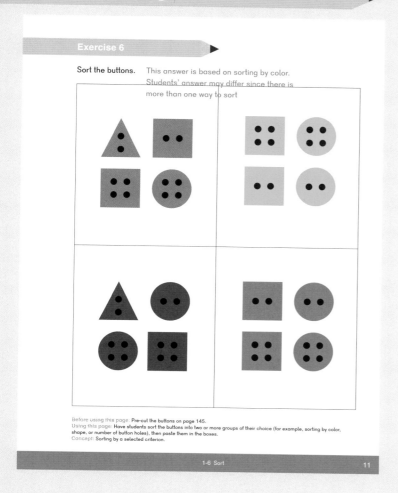

Before using this page: Pre-cut the buttons on page 145.
Using this page: Have students sort the buttons into two or more groups of their choice (for example, sorting by color, shape, or number of button holes), then paste them in the boxes.
Concept: Sorting by a selected criterion.

1-6 Sort   11

Sort the flowers.

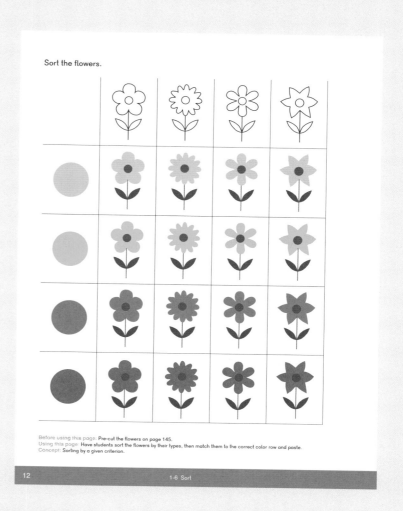

Before using this page: Pre-cut the flowers on page 145.
Using this page: Have students sort the flowers by their types, then match them to the correct color row and paste.
Concept: Sorting by a given criterion.

12   1-6 Sort

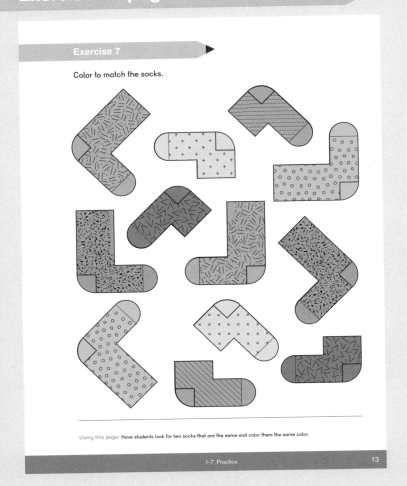

Exercise 7

Color to match the socks.

Using this page: **Have students look for two socks that are the same and color them the same color.**

1-7 Practice

13

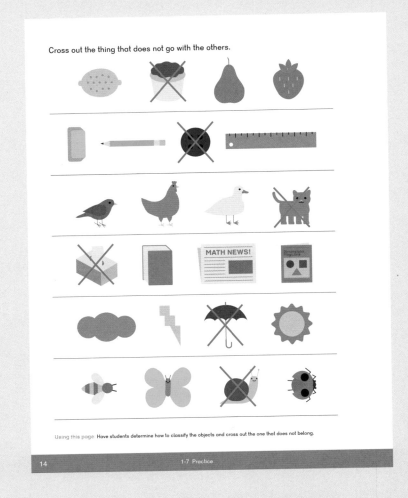

Cross out the thing that does not go with the others.

Using this page: **Have students determine how to classify the objects and cross out the one that does not belong.**

14

1-7 Practice

Teacher's Guide KA Chapter 1

Suggested number of class periods: 12 – 13

Lesson		Page	Resources		Objectives
	Chapter Opener	p. 27	TB:	p. 15	
1	Count to 5	p. 28	TB: WB:	p. 16 p. 15	Count from 1 to 5 and 5 to 1.
2	Count Things Up to 5	p. 31	TB: WB:	p. 19 p. 17	Recognize the quantity of objects by counting up to 5.
3	Recognize the Numbers 1 to 3	p. 34	TB: WB:	p. 22 p. 19	Recognize the numerals 1 to 3.
4	Recognize the Numbers 4 and 5	p. 37	TB: WB:	p. 25 p. 21	Recognize the numerals 4 and 5.
5	Count and Match	p. 39	TB: WB:	p. 27 p. 23	Count with cardinality and recognize numerals 1 to 5.
6	Write the Numbers 1 and 2	p. 41	TB: WB:	p. 29 p. 27	Write the numerals 1 and 2.
7	Write the Number 3	p. 44	TB: WB:	p. 32 p. 29	Write the numerals 1 to 3.
8	Write the Number 4	p. 46	TB: WB:	p. 34 p. 31	Write the numerals 1 to 4.
9	Trace and Write 1 to 5	p. 48	TB: WB:	p. 36 p. 33	Write the numerals 1 to 5.
10	Zero	p. 50	TB: WB:	p. 38 p. 35	Understand that an empty set is a set of zero items. Write numerals 0 to 5.
11	Picture Graphs	p. 53	TB: WB:	p. 42 p. 37	Read, interpret data, and create picture graphs.
12	Practice	p. 55	TB: WB:	p. 45 p. 39	Practice concepts from the chapter.
	Workbook Solutions	p. 57			

This chapter teaches sequential counting to and from 5, one-to-one correspondence of items to numbers up to 5, and numeral recognition and writing the numerals 0 to 5.

As children progress developmentally, they increase their knowledge of what a number is. Many students in Kindergarten can recite their number words in order. This rote counting evolves over time into the understanding that the numbers tell the count of items.

In **Dimensions Math® Kindergarten A**, students will learn to count and recognize the total quantity of items as the last number of their count. Additionally, the total quantity of items does not change even if they are rearranged. Mastery of counting skills is reached when students have automatic recognition of how many items they see (for small numbers), and when they recognize that rearrangement does not change the count.

Teaching the skill of one-to-one correspondence is also important. Some children recite numbers at a rate that doesn't match their finger pointing; they count to 8, while their fingers only touched 5 objects. Students demonstrate one-to-one correspondence by counting and naming objects one time and recognizing that the number they are saying corresponds to the number of items they have counted so far.

Particular attention should be paid to how students touch items they count. They should be asked to move items from the uncounted group to the counted group (not just touch) their small items. Encourage students to organize or line up the items as they are counted.

Students will practice recognizing that the numeral 5 refers to the quantity of 5 items. Students will begin by reciting numbers to 5, then counting objects, and then using five-frame cards to represent the count of objects.

Games, songs, and activities should be used and repeated as often as necessary. As students practice counting with increasingly abstract representations, they will begin to understand numbers.

Finally, picture graphs are introduced in **Chapter 2: Lesson 11** as a way of sorting and organizing data to make counting easier.

Storybooks that go along with content in this chapter can be added as time allows or for additional exploration. Storybook suggestions are listed on page 26 of this Teacher's Guide.

A Note on Writing Numerals

Numeral recognition and writing are introduced here to connect these symbols with the quantities being taught. This emphasis is not on the form of the numerals, but on students mastering how these symbols stand for the numbers of objects. Activities to promote the fine motor skills needed to properly print numbers are emphasized.

Just as the alphabet is the set of letters, written numbers are made up of a set of numerals. A numeral refers to the written symbol for 0 through 9, and number is the quantity. At this age, it is acceptable to use the term "number" in both cases.

It is expected that each school has a specific writing program to teach students numeral writing. Because programs vary in methods, **Dimensions Math®
Kindergarten** has a minimal set of instructions. In this chapter and **Chapter 3: Numbers to 10**, written numerals in the textbook are centered on a dotted line and do not touch the top and bottom guidelines.

For classrooms in which students are already counting with one-to-one correspondence to five and a systematic handwriting program is in place, the following alternate lesson order is suggested:

Lesson Order:

2-1: Count to 5
2-2: Count Things Up to 5
2-3: Recognize the Numbers 1 to 3 with Lessons:
  • 2-6: Write the Numbers 1 and 2
  • 2-7: Write the Number 3
2-4: Recognize the Numbers 4 and 5 with Lessons:
  • 2-8: Write the Number 4
  • 2-9: Trace and Write 1 to 5
2-10: Zero
2-11: Picture Graphs
2-12: Practice

Lessons on handwriting the numerals 1 through 5 do not have extensions, as it is expected that there will be additional practice on handwriting.

Materials

- Linking cubes
- Beans, beads, bear (or other animal) counters

Blackline Masters

- My Book of Numbers: Note that once a page for a specific number has been given to students, they will work on that page for multiple lessons. Therefore, new print-outs of My Book of Numbers pages are not required for each lesson.
- Blank Five-frame
- Five-frame Cards
- Number Cards
- Number Cards — Large
- Number Word Cards
- Picture Cards
- Blank Graph
- Fun Font Number Cards
- Draw and Cover Game Board
- Which One Doesn't Belong — Dice

Tactile methods of writing numerals

- Sandpaper: Cut numerals out of sandpaper with a die-cut machine or by hand and glue them to cardstock. Students can either cover the numerals with paper and make numeral rubbings or just trace the numerals with their fingers.
- Whiteboards: Have students write the numerals, then use their index finger to erase.
- Writing in a shallow pan:
  - Filled with sand, rice, or salt
  - Shaving cream writing
  - Gelatin or pudding
- Finger paint writing
- Wikki Stix

Storybooks

- *The Very Hungry Caterpillar* by Eric Carle
- *Five Little Monkeys Jumping on the Bed* by Eileen Christelow
- *Five Little Ducklings* by Sally Hopgood
- *The Five Chinese Brothers* by Claire Huchet Bishop
- *Pete the Cat and His Four Groovy Buttons* by James Dean and Eric Litwin
- *One Fish Two Fish Red Fish Blue Fish* by Dr. Seuss
- *The Five Senses* by Hervé Tullet

Letters Home

- Chapter 2 Letter

Chapter Opener

Lesson Materials

- Construction paper

This **Chapter Opener** is designed to introduce students to both the written numbers and objects that can be said to have five: fingers, toes, points on a star, and flower petals.

Explore

Have students discuss what they see on page 15. Ask students what they know about the numbers. Use this opportunity to pre-assess students' prior knowledge about numbers 1 to 5.

My Book of Numbers: Begin the chapter with this project that will continue through the lessons. Students will be drawing pictures of a quantity of objects, adding pre-printed five-frame representations, then adding written numerals to the pages.

For the **Chapter Opener**, have students trace their hands on sheets of construction paper to create covers for their books (they may work with partners). Include their names and book titles.

Chapter 2

Numbers to 5

15

Extend

▲ **Other Numbers to 5**

Introduce students to counting in other representations or languages, such as:

Representation systems:

- Tally marks; I, II, III, IIII, IIII
- Roman numerals: I, II, III, IV, V
- Chinese characters: 一, 二, 三, 四, 五

Counting words in various languages:

- Spanish: Uno, dos, tres, cuatro, cinco
- German: Eins, zwei, drei, vier, fünf
- Chinese pronunciation: yi, er, san, si, wu

Objective

- Count from 1 to 5 and 5 to 1.

Lesson Materials

- Counters or linking cubes, 5 per student

Explore

Provide each student with 5 counters. Ask them how many counters they have. Note if any students touch or move the same counters more than once and arrive at an incorrect answer. As students count, encourage them to move the counters from the uncounted group to the counted group.

Have students count along with Alex from 1 to 5 and 5 to 1 on page 16.

Learn

Have students discuss the beach scene on page 17. As they notice numbers of objects, ask them to show the same number with their counters. For example:

- I see 4 shovels. (Have all students count out and show 4 counters.)
- I see 5 sand dollars. (Have all students count out and show 5 counters.)

Students with advanced number sense may see and count quantities greater than 5, such as 6 legs on seagulls or 25 points on the sand dollars. Provide these students additional counters and challenge them to organize their counters into groups.

Have students represent the toys with counters as they touch and count Mei's toys on page 18.

Whole Group Activities

▲ **Five Little Monkeys**

Materials: Five Little Monkeys Jumping on the Bed Video (VR).

Recite Five Little Monkeys Jumping on the Bed.

▲ **I Spy**

Play with numbers to 5 in the classroom.

▲ **Finger Flash**

Show students 1 to 5 fingers for long enough to see but not count (try two seconds). See if they know how many fingers were shown.

Small Group Activities

▲ **Textbook** Page 18

▲ **Count and Sort**

Materials: Sorting mats

Using sorting mats (color, size, shape, etc.), students find 5 objects for each mat.

▲ **How Many Can You Make?**

Materials: Squares, cubes

Using 5 squares or cubes, students make as many different configurations as possible.

▲ **My Book of Numbers**

Materials: My Book of Numbers (BLM) pages 1 – 5, small objects that can be glued to paper

Give each student Book of Numbers (BLM) pages 1 – 5. This is the first time students will work with their pages, so let them be creative as they draw pictures on their page to represent the numbers 1, 2, 3, 4, and 5, then have them glue objects to each page for the number shown on that page.

In later lessons, they will practice writing the numerals being explored in the lesson on their pages. Give students additional pages for each number as needed.

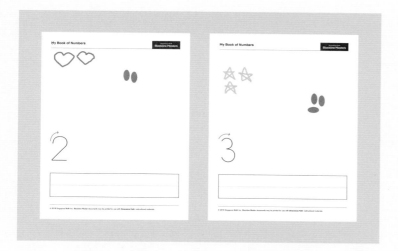

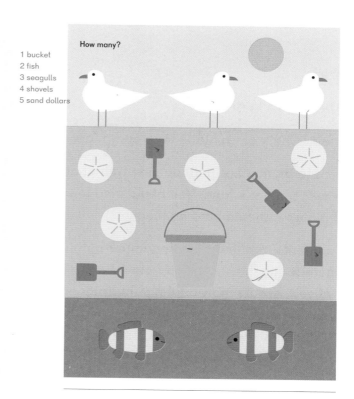

How many?

1 bucket
2 fish
3 seagulls
4 shovels
5 sand dollars

Objective: **Count to 5.**

2-1 Count to 5   17

17

Extend

★ Number Path Walk

Materials: Chalk or painter's tape and paper plates

Create a large number path outside with chalk, or inside with painter's tape. You could also tape down paper plates with numbers on them.

Have students hop through the number path saying each number as they hop on it from 1 to 5 and 5 to 1. Alternatively, create individual number paths. Call a number and have students hop to the number on their path. Call objects with quantities, such as, "Hop to the number of wheels a tricycle has," or, "Hop to the number of legs a dog has."

| Home | 1 | 2 | 3 | 4 | 5 |

Objective: Count to 5.

18   2-1 Count to 5

Lesson 2 Count Things Up to 5

Objective

- Recognize the quantity of objects by counting up to 5.

Lesson Materials

- Blank Five-frames (BLM), 1 per student
- Counters or other small objects, 5 for each pair of students

Explore

Have students discuss how many of each type of object they see in the bedroom on page 19.

Learn

Give each student 5 small objects and a Blank Five-frame (BLM), and have them put their objects on the cards and count.

Ask students to represent objects from the illustration on page 19 on their Blank Five-frames (BLM). Examples:

- I see 4 books. (Have all students make 4 on their blank five-frame.)
- I see 3 socks. (Have all students make 3 on their blank five-frame.)

If needed, prompt students with questions such as, "I see 4 of something. What do you think it is?"

Lesson 2
Count Things Up to 5 ②

Look and talk.
How many?

19

Objective: Create sets and count the number of objects in the set.

2-2 Count Things Up to 5   19

Have students look at page 20 and the way Dion organized his fruit. Ask them to discuss what they notice.

Whole Group Activities

▲ I Spy

Play with numbers to 5 in the classroom.

▲ Finger Flash

Show students 1 to 5 fingers for long enough to see but not count (try two seconds). See if they know how many fingers were shown.

▲ Five-frame Flash

Materials: Five-frame Cards (BLM) 1 to 5

Show students Five-frame Cards (BLM) 1 to 5. See if they can say how many dots were shown on the card. To extend this activity, see if they can say how many blank spaces were shown on the card.

Small Group Activities

▲ Textbook Page 21

Have students show the quantity of objects in the classroom on a Blank Five-frame (BLM). Examples:

- I see 2 tables. (The student should show 2 counters on her five-frame.)
- I see 4 windows (or panes). (The student should show 4 counters on his five-frame.)

▲ Count and Sort

Materials: Sorting mats

Using sorting mats (color, size, shape, etc.), students find 5 objects for each mat.

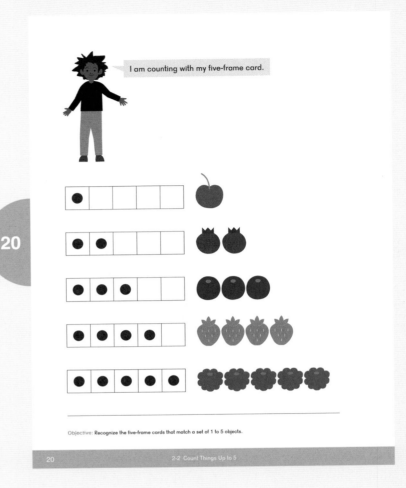

I am counting with my five-frame card.

20

Objective: Recognize the five-frame cards that match a set of 1 to 5 objects.

20   2-2 Count Things Up to 5

▲ How Many Can You Make?

Materials: Squares, cubes

Using 5 squares or cubes, students make as many different configurations as possible.

▲ Bead Bracelets

Materials: String, 5 beads per student

Make bead bracelets with 5 beads. Students can wear to touch and count.

▲ My Book of Numbers

Materials: My Book of Numbers (BLM) pages 1–5, Blank Five-frames (BLM)

Have students fill in a Blank Five-frame (BLM) to represent each number, 1 to 5, and glue them to the correct My Book of Numbers (BLM) page below their pictures.

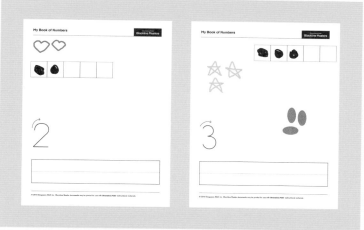

Exercise 2 • page 17

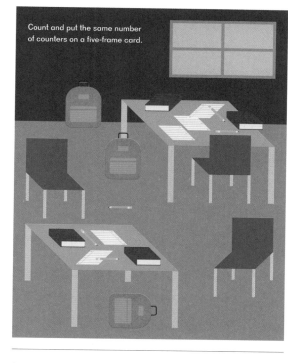

Extend

★ Five-frame Fill-up

Materials: Colored counters, Blank Five-frames (BLM), die

Players choose one color for their counter. They take turns rolling the die and placing their color counters on any Blank Five-Frame (BLM) to match the roll. When a player fills a five-frame with an exact roll, they score a point. The player with the most points is the winner.

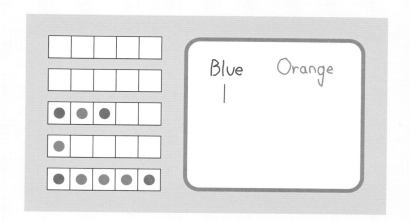

Objective

- Recognize the numerals 1 to 3.

Lesson Materials

- Sets of cards, enough for each student to have 1 card each of the following:
  - Number Cards (BLM) 1 to 3
  - Picture Cards (BLM) 1 to 3
  - Five-frame Cards (BLM) 1 to 3
- Fun Font Number Cards (BLM) 1 to 3

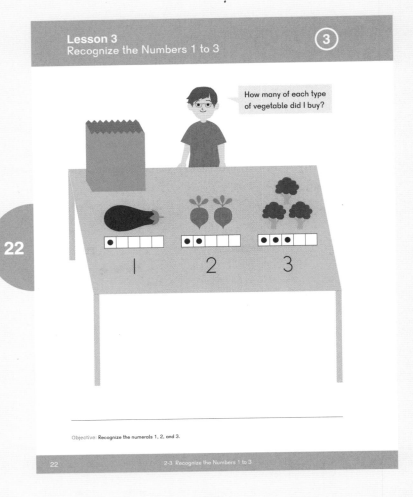

Explore

Pass out Picture Cards (BLM) 1 to 3 and Five-frame Cards (BLM) 1 to 3 randomly to students so each student has 1 card.

Have students mingle and find other students with the same number represented either by picture or five-frame to form three groups. Ask students to explain why they belong in the group they chose.

While the students are grouped, show them the Number Card (BLM) that matched the quantity on the group's cards. Tell them that this is the number 1, 2, etc.

Show students Fun Font Number Cards (BLM) 1 to 3 and discuss how the written number is the same or similar.

Exchange the Picture Cards (BLM) for Number Cards (BLM) and have students find their group with the new cards. Repeat the activity until students can identify the different representations of the numbers 1, 2, and 3.

Learn

Discuss Emma's vegetables and the five-frame representations. Have students show the number card for the eggplant, then the radish, then the broccoli.

Whole Group Activities

▲ Follow Me

Materials: Number Cards (BLM) 1 to 3

Students take turns drawing a Number Card (BLM) and choosing an action (hopping, clapping, stomping, etc.). For example, the 2 card is drawn and the student chooses to hop. All students hop 2 times. Play continues with the next student drawing a new number card and choosing an action.

▲ Count and Sit

All students start by standing in a circle and counting off, "1, 2, 3." The student that says, "3," sits. The count continues around the circle. Only students standing count. The activity ends when only one student is remaining standing.

Small Group Activities

▲ **Textbook** Pages 23 and 24

▲ **How Many in Here?**

Materials: 3 linking cubes, bag, Number Cards (BLM)

Players take turns being the Counter. While the Counter has his eyes closed, his partner places up to 3 linking cubes in an opaque bag. With his eyes still closed, the Counter feels and counts the cubes in the bag. When he has counted the cubes, he opens his eyes and finds that number on a Number Card (BLM).

▲ **My Book of Numbers**

Materials: My Book of Numbers (BLM) pages 1 – 3, Fun Font Number Cards (BLM)

Have students cut out Fun Font Number Cards (BLM) 1 to 3 and glue them to the correct My Book of Numbers (BLM) page next to the five-frame.

Draw circles to match each number.

Objective: Recognize the numerals 1, 2, and 3.

2-3 Recognize the Numbers 1 to 3   23

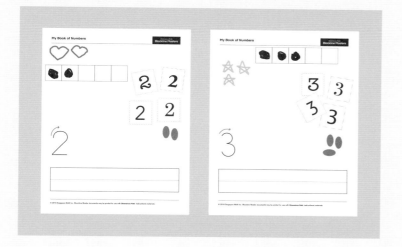

Take it Outside

▲ Number Tag

Materials: Number Cards — Large (BLM) 1 to 3

Set up 3 bases outside and post a Number Card — Large (BLM) at each base. Choose a student to be Catcher. Have the other students line up on a start line.

Call out a number from those used at the bases. All students will try to reach the base where that number is posted before being tagged by the Catcher. Extend with more complicated directions like, "1 or 3," "2 or 1," "Any base but 3," or, "The number after 1."

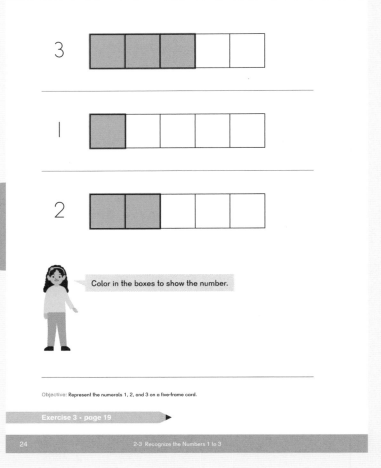

Color in the boxes to show the number.

Objective: Represent the numerals 1, 2, and 3 on a five-frame card.

Exercise 3 • page 19

24   2-3 Recognize the Numbers 1 to 3

Exercise 3 • page 19

Extend

★ Mirror

Materials: Colored square tiles, 1 each of three different colors per student

Using three differently colored square tiles, Player 1 creates a design without showing it to Player 2. Player 1 then describes the design to Player 2, who tries to create the same design based on Player 1's description.

Objective

- Recognize the numerals 4 and 5.

Lesson Materials

- Sets of cards, enough for each student to
  have 1 card each of the following:
  - Number Cards (BLM) 1 to 5
  - Picture Cards (BLM) 1 to 5
  - Five-frame Cards (BLM) 1 to 5
  - Fun Font Number Cards (BLM) 1 to 5

Lesson 4
Recognize the Numbers 4 and 5
④

Look and talk.
Which groups have 4?
Which groups have 5?

25

Objective: Count and match a set of 4 or 5 objects to the numerals.

2-4 Recognize the Numbers 4 and 5   25

Explore

Pass out Picture Cards (BLM) 1 to 5 and Five-frame
Cards (BLM) 1 to 5 randomly to students so each
student has 1 card.

Have students mingle to form five groups by finding
other students with the same number represented
either by Picture Card (BLM) or Five-frame Card
(BLM). Ask students to explain why they belong in
the group they chose.

While the students are grouped, show them the
Number Card (BLM) that matches the quantity on the
group's cards. For example, show the card for 4 and
tell them, "This is the number 4."

Show students Fun Font Number Cards (BLM) and
discuss how the written number is the same or
similar. Collect the Picture Cards (BLM), hand out
Number Cards (BLM), and repeat the activity.

Learn

Discuss the forest scene on page 25. Have students
share what they see and show the corresponding
Number Card (BLM) and/or Five-frame Card (BLM).
For example:

- I see 4 bears. (Have students show either the
  4 Number Card (BLM) or the Five-frame Card
  (BLM) for four.)

- I see 5 petals on each flower. (Have students
  show either the Number Card (BLM) or the Five-
  frame Card (BLM) for 5.)

If needed, prompt student with questions such as,
"I see 5 of something. What do you think it is?"

Whole Group Activities

▲ Group Up

Play music and have students walk around the
classroom. Stop the music and hold up a number
card. Students must get into groups of the chosen
number as quickly as possible.

▲ Count and Sit

Play **Count and Sit** as described in **Lesson 3**, but
this time, have standing students count off by ones to
5. The student that says, "5," sits.

Small Group Activities

▲ **Textbook** Page 26

▲ **How Many in Here?**

Materials: 5 linking cubes, bag, Number Cards (BLM)

Play as described in **Lesson 3**, but this time, use up to 5 linking cubes.

▲ **Draw and Cover**

Materials: Picture Cards (BLM) 1 to 5 or Five-Frame Cards (BLM) 1 to 5, Draw and Cover Game Board (BLM)

Use a deck of Picture Cards (BLM) or Five-Frame Cards (BLM) 1 to 5 and the Draw and Cover Game Board (BLM). Students take turns drawing a picture card and covering the corresponding number on the game board. The first player to cover all numbers wins.

▲ **My Book of Numbers**

Materials: My Book of Numbers (BLM) pages 4 – 5, Fun Font Number Cards (BLM)

Have students cut out Fun Font Number Cards (BLM) 4 to 5 and glue them to the correct Book of Numbers (BLM) page next to the five-frame.

Take It Outside

▲ **Number Tag**

Materials: Number Cards — Large (BLM)

Set up 5 bases outside and play number tag.

▲ **Four Square**

▲ **Hopscotch to 5**

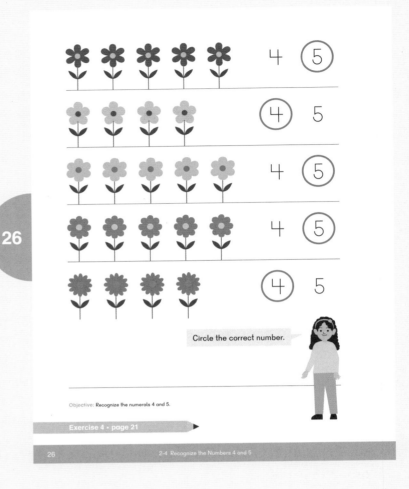

Circle the correct number.

Objective: Recognize the numerals 4 and 5.

Exercise 4 • page 21

26   2-4 Recognize the Numbers 4 and 5

▲ **Number Hop**

On the playground, have each student draw a large five-frame with 5 dots filled in with chalk. Call numbers. On each call, have students touch the correct number of dots on the five-frame according to this key:

1 — both feet on one dot
2 — each foot on a dot
3 — each foot and one hand on a dot
4 — each foot and each hand on a dot
5 — have students use their heads

Exercise 4 • page 21

Lesson 5 Count and Match

Objective

- Count with cardinality and recognize numerals 1 to 5.

Lesson Materials

- Five-frame Cards (BLM) 1 to 5, or Picture Cards (BLM) 1 to 5
- Number Cards (BLM) 1 to 5, 1 card per student

Explore

Give each student one Number Card (BLM) 1 to 5. Hold up a Picture Card (BLM) 1 to 5 or Five-frame Card (BLM) 1 to 5. Say, "I have this many, who has the number?"

Students with the corresponding numeral will hold up their Number Card (BLM). Repeat with all numbers, in random order. Ask students to organize themselves silently into groups so that each group contains the numerals 1 to 5. If a group already has a 2, the student finds a different group that needs a 2. Once groups are assembled, have students line up in order from 1 to 5.

Learn

Give pairs of students a set of Number Cards (BLM) 1 to 5. Students place the correct amount of counters on the corresponding Number Card (BLM).

Whole Group Activities

▲ **Listen and Count**

Materials: Number Cards (BLM) 1 to 5

Give each student a set of Number Cards (BLM) 1 to 5. Designate a Leader and have him clap 1 to 5 times. Have other students listen and count silently and represent the number of claps heard by holding up a number card.

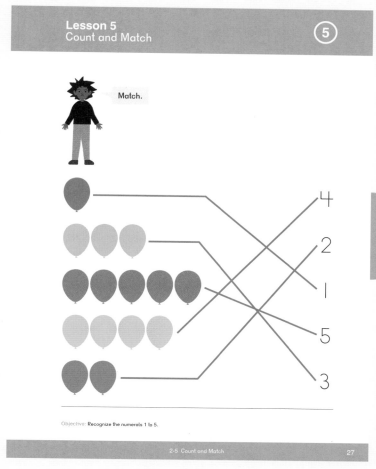

Match.

Objective: Recognize the numerals 1 to 5.

2-5 Count and Match   27

▲ Follow Me

Materials: Number Cards (BLM) 1 to 5

Students take turns drawing a Number Card (BLM) and choosing an action (hopping, clapping, stomping, etc.), as introduced in Lesson 3.

Small Group Activities

▲ **Textbook** Pages 27 and 28

▲ Simon Says

Materials: Number Cards (BLM) 1 to 5

Students take turns being Simon. Simon draws a Number Card (BLM) and gives commands such as jump, clap, stomp, or spin a designated number of times. Instead of saying the number, Simon will hold up that number card.

▲ My Book of Numbers

Students that need to finish up any of their pages 1 to 5 should do so during this lesson. By the end of this lesson, all student pages should have the following for each numeral 1 to 5:

- Picture
- Glued objects
- Five-frame Card (BLM)
- Fun Font Number Card (BLM)

Students will complete their books by writing the numeral in the next set of lessons.

Take it Outside

▲ Nature Hike

Have students collect 1, 2, 3, 4, and 5 similar items, such as 1 acorn, 2 leaves, 3 pine needles, 4 pebbles, and 5 sticks, etc. (Could be added to My Book of Numbers.)

▲ Jump the Number

Materials: Jump rope

Have two students twirl each jump rope. The rest of the students line up. Have the first student in each line stand next to a jump rope, then one of the twirlers calls a number as specified in the lesson. When the jumper has jumped the same number of times called out by the twirler, that jumper jumps out and goes to the end of the line. Continue until all students have had a chance to be the jumper.

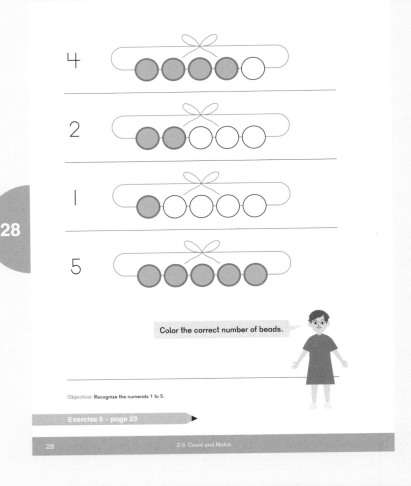

Color the correct number of beads.

Objective: Recognize the numerals 1 to 5.

Exercise 5 • page 23

28   2-5 Count and Match

◀ Exercise 5 • page 23

Extend

★ What's Missing?

Materials: Five-frame Cards (BLM) 1 to 5 or Number Cards (BLM) 1 to 5

Partners use one set of either Five-frame Cards 1 to 5 (for beginners) or Number Cards 1 to 5 (for an extension). Player 1 removes one of the cards from the set and hands the cards to Player 2. Player 2 puts the cards in order and figures out the missing card. Players then switch roles.

Objective

- Write the numerals 1 and 2.

Learn

Students will be writing numerals in this and the next three lessons. Students should be at desks or tables to encourage proper posture and hand grip. Beginning with textbook page 29, have students trace the shapes, then write the numerals, always starting at the top.

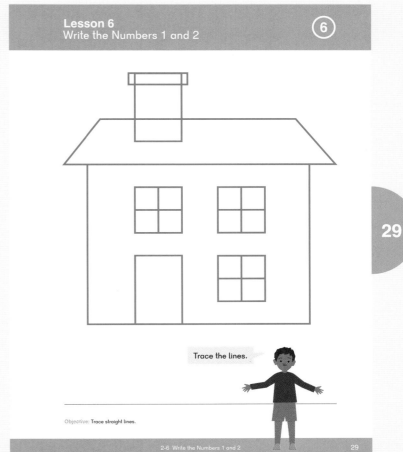

Have students trace first with their index finger, then use a pencil to write the numerals.

Students may recite as they write 1s on page 30:

I start at the top and draw a straight line.
My number 1 is mighty fine.

Students may recite as they write 2s on page 31:

A curve and a line will be the clue,
that is how I make the number 2.

Whole Group Activity

▲ Conducting

Materials: Stick or straw

Have students use a small stick or straw and become conductors of their number orchestra. Each student forms the numerals in the air.

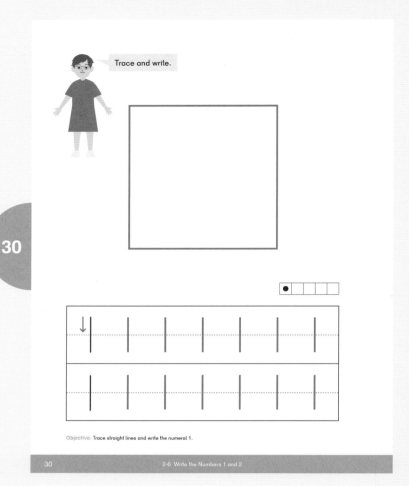

Trace and write.

Objective: Trace straight lines and write the numeral 1.

30   2-6 Write the Numbers 1 and 2

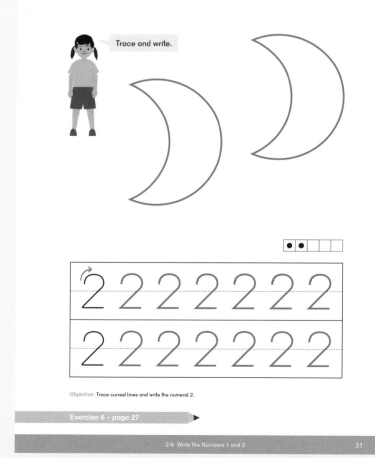

Trace and write.

Objective: Trace curved lines and write the numeral 2.

Exercise 6 • page 27

2-6 Write the Numbers 1 and 2   31

Small Group Activities

▲ Textbook Pages 30 and 31

Centers should include counting small objects to develop fine motor skills and writing with fingers or pencil. Some ideas:

▲ Feeling for 5

Materials: Containers of rice or sand, toys, counters and/or beans, paper cup

Set up work stations with containers of rice or sand in which you have buried small toys, counters, and/or beans. Students each remove 5 objects and puts them in a paper cup. Save this center for students to use in the following lesson.

▲ Tactile Writing

Set up a center where students can practice writing numerals in a variety of methods such as:

- Sandpaper: Cut numerals out of sandpaper with a die-cut machine or by hand. Glue the numerals to cardstock paper. Students can either cover the numerals with paper and make numeral rubbings or just trace the numerals with their fingers.
- Whiteboards: Have students write the numerals, then use their index finger to erase.
- Writing in a shallow pan:
  - Filled with sand, rice, or salt
  - Shaving cream writing
  - Gelatin or pudding
- Finger paint
- Wikki Stix

▲ My Book of Numbers

Materials: My Book of Numbers (BLM) pages 1 – 2

Have students write numerals 1 and 2 on the corresponding My Book of Numbers (BLM) pages.

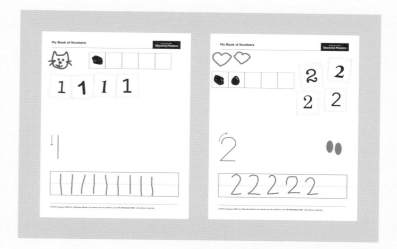

Exercise 6 • page 27

Objective

- Write the numerals 1 to 3.

Learn

Students should be at desks or tables to encourage proper posture and hand grip. Beginning on page 32, have students trace the shamrock before writing the numerals, first tracing with a finger then writing with a pencil. Students may recite as they write:

Start at the top
Curve out and in, you'll see.
Do it all again.
You've made a number 3.

Whole Group Activity

▲ **Conducting**

Materials: Stick or straw

Have students use a small stick or straw and become conductors of their number orchestra. Each student forms the numerals in the air.

Small Group Activities

▲ **Textbook** Page 33

Centers should include counting small objects to develop fine motor skills and writing with fingers or pencil. Some ideas:

▲ **Feeling for 5**

Materials: Containers of rice or sand, toys, counters and/or beans, paper cup

Repeat the activity as introduced in the previous lesson.

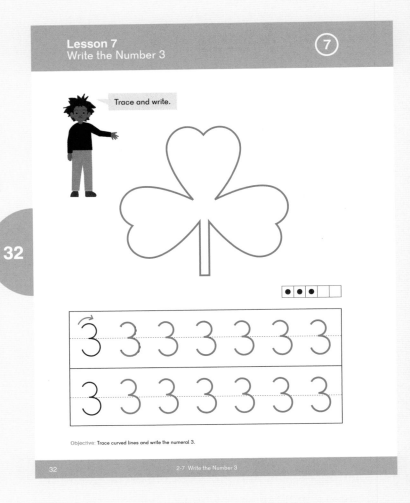

Lesson 7
Write the Number 3 ⑦

Trace and write.

Objective: Trace curved lines and write the numeral 3.

32   2-7 Write the Number 3

▲ Play Dough Numbers

Materials: Play dough

Have students make the numerals out of dough.

▲ Sandpaper Numbers

Cut the numerals out of sandpaper before class. Have students trace sandpaper numerals with their index fingers.

▲ Sand Writing

Materials: Sand, low-rimmed pans

Spread a layer of sand on a low-rimmed pan, and have students write the numerals with their fingers.

▲ My Book of Numbers

Materials: My Book of Numbers (BLM) page 3

Have students write the numeral 3 on the corresponding My Book of Numbers (BLM) page.

Take it Outside

▲ Walk and Paint the Numbers

Materials: Sidewalk chalk, paintbrushes, cups of water

Use chalk to write very large numerals 1, 2, and 3 (about 4 feet tall) outside, each in a different color. Surrounding the large numerals, write the same numerals, but smaller (about 8 inches tall). Students line up single file and walk over the numerals, starting at the top, while saying the rhymes. Provide students with a cup of water and a small paintbrush to trace over the smaller numerals.

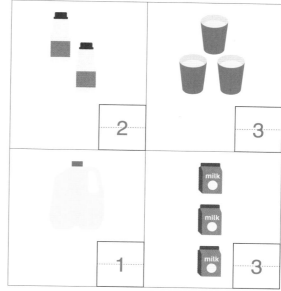

33

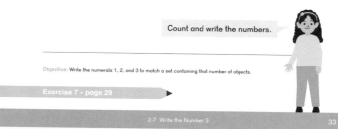

Count and write the numbers.

Objective: Write the numerals 1, 2, and 3 to match a set containing that number of objects.

Exercise 7 • page 29

2-7 Write the Number 3   33

Exercise 7 • page 29

Teacher's Guide KA Chapter 2   45

Objective

- Write the numerals 1 to 4.

Learn

Students should be at desks or tables to encourage proper posture and hand grip. Students may recite as they write:

Short line down, then to the right,
A long line down, my 4 is just right.

Whole Group Activity

▲ **Partner Numbers**

Materials: Number Cards — Large (BLM) 1 to 4

Play music, and when the music stops, hold up a Number Card — Large (BLM) 1 to 4. Students use their bodies to form the numerals with a partner or partners.

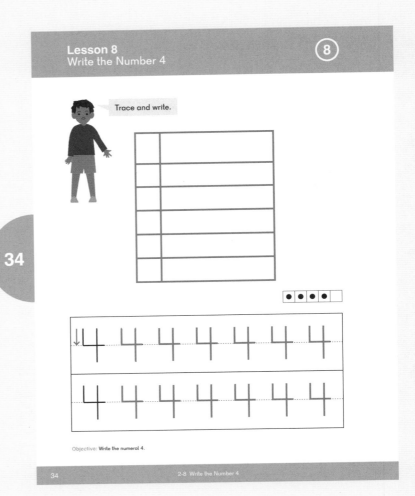

Small Group Activities

▲ **Textbook** Page 35

Centers should include counting small objects to develop fine motor skills and writing with fingers or pencils.

▲ **Breath Art**

Materials: Eye dropper or small spoons, paint, construction paper

Have students use an eye dropper or a tiny spoon to put paint on construction paper. Then have them blow through a straw to move the paint on the paper. After allowing the students to create their own masterpieces, ask them to try to make a 1 and a 4 by blowing paint on a piece of paper.

▲ **Geoboard 4**

Materials: Geoboard, rubber bands

Have students create as many 4s on the geoboard as they can.

▲ **My Book of Numbers**

Materials: My Book of Numbers (BLM) page 4

Have students write the numeral 4 on the corresponding My Book of Numbers (BLM) page.

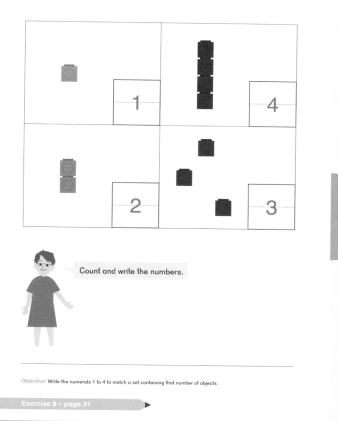

Count and write the numbers.

Objective: Write the numerals 1 to 4 to match a set containing that number of objects.

Exercise 8 • page 31

2-8 Write the Number 4   35

Take it Outside

▲ **Walk and Paint the Numbers**

Materials: Sidewalk chalk, paintbrushes, cups of water

Use chalk to write very large numerals 1 – 4 (about 4 feet tall) outside, each in a different color. Surrounding the large numerals, write the same numerals, but smaller (about 8 inches tall). Students line up single file and walk over the numerals, starting at the top, while saying the rhymes. Provide students with a cup of water and a small paintbrush to trace over the smaller numerals.

Exercise 8 • page 31

Objective

- Write the numerals 1 to 5.

Learn

Students should be at desks or tables to encourage proper posture and hand grip. Beginning on page 36, have students trace the blank five-frames, then write the numerals, always starting at the top. Students may recite as they write:

First a line, then take a dive,
Around I go to make the number 5.

Whole Group Activity

▲ **Flash and Write**

Materials: Five-frame Cards (BLM) 1 to 5

Flash a Five-frame Card (BLM) and have students write the corresponding numeral on whiteboards. Clear boards and repeat.

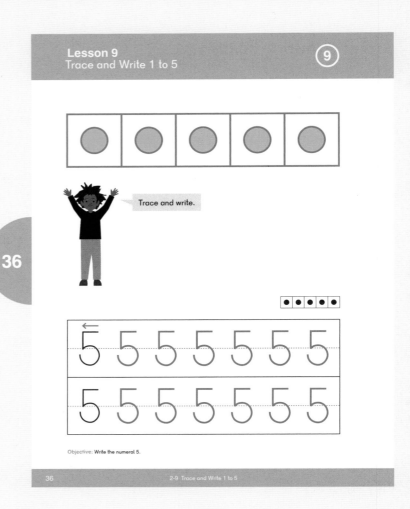

Lesson 9
Trace and Write 1 to 5
⑨

Trace and write.

5 5 5 5 5 5 5

5 5 5 5 5 5 5

Objective: Write the numeral 5.

36   2-9 Trace and Write 1 to 5

Small Group Activities

▲ **Textbook** Page 37

Centers should include counting small objects to develop fine motor skills and writing with fingers or pencils.

▲ **My Book of Numbers**

Materials: My Book of Numbers (BLM) page 5

Have students write the numeral 5 on the corresponding My Book of Numbers (BLM) page.

▲ **Punch Numbers**

Materials: Push-pin, golf tee, or hole punch, construction paper or index cards

Have students use a push-pin, golf tee, or hole punch to punch out the outline of the numerals designated in the lesson on small pieces of construction paper or index cards.

Take it Outside

▲ **Walk and Paint the Numbers**

Materials: Sidewalk chalk, paintbrushes, cups of water

Use chalk to write very large numerals 1 – 5 (about 4 feet tall) outside, each in a different color. Surrounding the large numerals, write the same numerals, but smaller (about 8 inches tall). Students line up single file and walk over the numerals, starting at the top, while saying the rhymes. Provide students with a cup of water and a small paintbrush to trace over the smaller numerals.

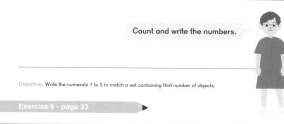

Count and write the numbers.

Objective: Write the numerals 1 to 5 to match a set containing that number of objects.

Exercise 9 • page 33

2-9 Trace and Write 1 to 5   37

Exercise 9 • page 33

Lesson 10 Zero

Objectives

- Understand that an empty set is a set of zero items.
- Write numerals 0 to 5.

Lesson Materials

- Five-frame Card (BLM) 0
- Number Card (BLM) 0

Explore

Ask students questions to which the answers will be, "None." Examples:

- How many live elephants are in our class today?
- How many students drove themselves to school in a car today?
- What else is there none of in the class today?

Tell students that there is a number for "none" or "nothing." Show them a Five-frame Card (BLM) with none of the dots filled in and a Number Card (BLM) for 0. Introduce the number 0.

Have students count back from 5 to 0 and jump in the air while saying, "5... 4... 3... 2... 1... 0... blast off!"

Have students make a giant circle in the air with their fingers.

Learn

On page 38, discuss Dion's puppy, Spot, and the five-frame representation for 0. Give students counters and have them act out the scenario of the biscuits with their counters.

On page 39, have students trace the egg first and then the number 0 with their index finger, then use a pencil to write the number. Students may recite as they write:

0 looks like a racing track
Go around and come right back.

Teacher's Guide KA Chapter 2   © 2017 Singapore Math Inc.

Have students practice writing the numerals 0 to 5 on pages 40 and 41.

Whole Group Activity

▲ Conducting

Materials: Stick or straw

Have students use a small stick or straw and become conductors of their number orchestra. Each student forms the numerals in the air.

Small Group Activities

▲ Textbook Pages 40 – 41

▲ My Book of Numbers

Materials: My Book of Numbers (BLM) page 0, Blank Five-frame (BLM), Number Card (BLM) 0

Have students add a Blank Five-frame (BLM), Number Card (BLM), and written numeral for 0 to the My Book of Numbers (BLM) page for 0.

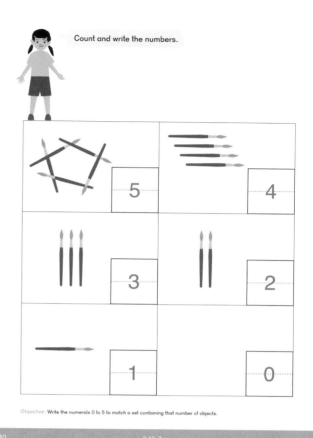

Count and write the numbers.

5	4
3	2
1	0

Objective: Write the numerals 0 to 5 to match a set containing that number of objects.

5	2
4	0
1	3

Count and write the numbers.

Objective: Write the numerals 0 to 5 to match a set containing that number of objects.

Exercise 10 • page 35

Take it Outside

▲ **Walk and Paint the Numbers**

Materials: Sidewalk chalk, paintbrushes, cups of water

Use chalk to write very large numerals 0 – 5 (about 4 feet tall) outside, each in a different color. Surrounding the large numerals, write the same numerals, but smaller (about 8 inches tall). Students line up single file and walk over the numerals, starting at the top, while saying the rhymes. Provide students with a cup of water and a small paintbrush to trace over the smaller numerals.

Exercise 10 • page 35

Extend

★ **Five Tops Game**

Materials: Classroom board game, Number Cards (BLM) 0 to 5 or a die with a 0 in place of the 6

Use any classroom game that moves with cards or dice, by using Number Cards (BLM) 0 to 5 or a die with a 0 sticker over the 6.

Play regular rules for the board game, but move with either cards or a modified die.

Lesson 11 Picture Graphs

Objective

- Read, interpret data, and create picture graphs.

Lesson Materials

- Linking cubes in 4 colors, 5 per student
- Blank Graph (BLM)

Prior to **Explore**, use tape to create a graph outline on the floor similar to the one on page 42, but with only three columns. Alternatively, create one in advance on large chart paper.

Explore

Provide students with up to 5 each of three different colors of cubes. Ask students how they could organize the cubes to make them easier to count by color. Have students share their ideas.

Show students the graph. Ask them how they could use the graph to organize the cubes. Have a few students take turns putting the cubes on the big graph and counting how many of each color there are. Have students suggest labels for the columns and a title for the graph (colors of cubes).

Students can put their cubes on a Blank Graph (BLM) and share how they organized the cubes.

Learn

Have students use the Blank Graph (BLM) and linking cubes to show the fruits on page 42.

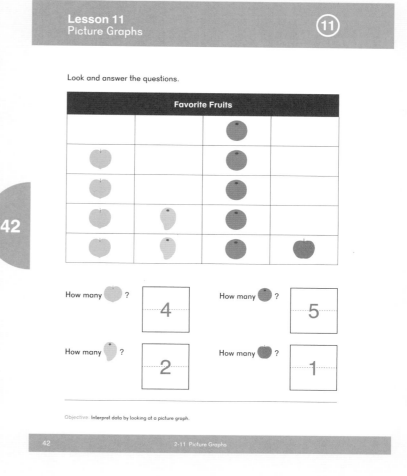

Whole Group Activity

▲ Shoe Sort

Materials: Painter's tape or butcher paper

Create a large graph on the floor with painter's tape or butcher paper and have some students take off one of their shoes. Have students discuss how to sort the shoes.

Students may choose to create a graph based on color, type of shoes, size, slip-on or laces, etc.

Small Group Activities

▲ **Textbook** Pages 42 – 44

Write the numerals.

▲ **Make a Graph**

Materials: Blank Graph (BLM), manipulatives, no more than 5 of each category

Have students sort the objects and draw pictures of the objects in the correct column on a Blank Graph (BLM). They can write the correct numeral below each column.

Exercise 11 • page 37

Extend

★ **Which One Doesn't Belong?**

Materials: Which One Doesn't Belong? – Dice (BLM)

Find a reason why one die does not belong.

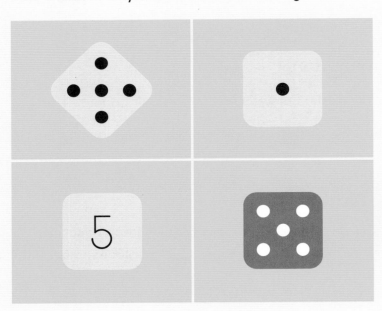

Possible student answers:

- One die is red, the rest are white.
- One die has a number, the rest have dots.
- One die shows 1, the rest show 5.
- One die is turned differently from the others.

Look and answer the questions.

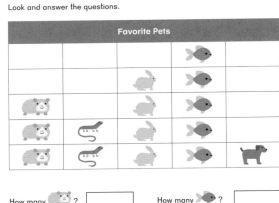

How many 🐹 ? `3`   How many 🐟 ? `5`

How many 🦎 ? `2`   How many 🐕 ? `1`

How many 🐰 ? `4`

Objective: Interpret data by looking at a picture graph.

2-11 Picture Graphs   43

Color the boxes to show how many.

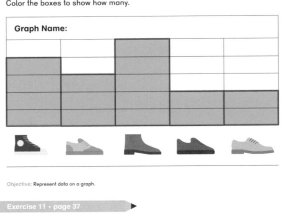

Objective: Represent data on a graph.

Exercise 11 • page 37

44   2-11 Picture Graphs

Lesson 12 Practice

Objective

- Practice concepts from the chapter.

After students complete the **Practice** in the textbook, have them continue counting and writing numerals from 0 to 5, using **Activities** from the chapter.

Whole Group Activity

▲ Silent Match 'Em Up

Materials: Number Cards (BLM) 0 to 5, Picture Cards (BLM) 0 to 5, Five-frame Cards (BLM) 0 to 5, Number Word Cards (BLM) 0 to 5

Pass out cards to students and have students find other students with cards that match their own card without speaking.

Students can also line up in order from 0 to 5 with students holding cards with the same number standing in front of or behind their number match. For example, these cards would all go together:

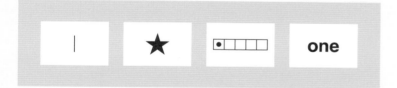

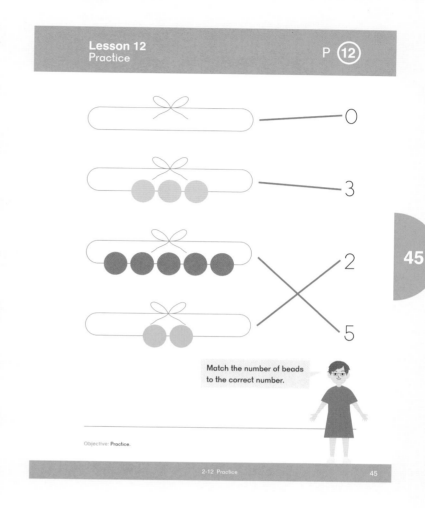

Small Group Activities

▲ Match

Materials: Number Cards (BLM) 0 to 5, Picture Cards (BLM) 0 to 5, Five-frame Cards (BLM) 0 to 5, Number Word Cards (BLM) 0 to 5

Students arrange the cards, faceup, in a grid. Students take turns finding two cards that go together.

★ Memory

Materials: Number Cards (BLM) 0 to 5, Picture Cards (BLM) 0 to 5, Five-frame Cards (BLM) 0 to 5, Number Word Cards (BLM) 0 to 5

Students arrange the cards, facedown, in a grid. Students take turns finding two cards that go together.

Exercise 12 • page 39

Extend

★ Make a Match

Materials: Picture Cards (BLM) 0 to 5, Number Cards (BLM) 0 to 5, Five-frame Cards (BLM) 0 to 5, Number Word Cards (BLM) 0 to 5

This can be a whole class activity that uses number word, numeral, and ten-frame representations. Pass cards to students and have them find cards that match their own card without speaking.

Students can also line up in order from 0 to 5.

For example, these cards would line up together:

1	one	●○○○○ / ○○○○○	🦕

Incorporate Number Word Cards (BLM) 0 to 5 for students who are identifying words or reading.

Write the numbers.

Objective: **Practice.**

2-12 Practice   47

47

Color in the picture graph.

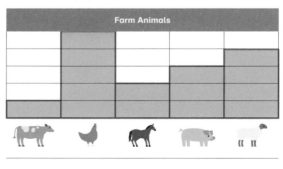

Farm Animals

Objective: **Practice.**

Exercise 12 • page 39

48   2-12 Practice

48

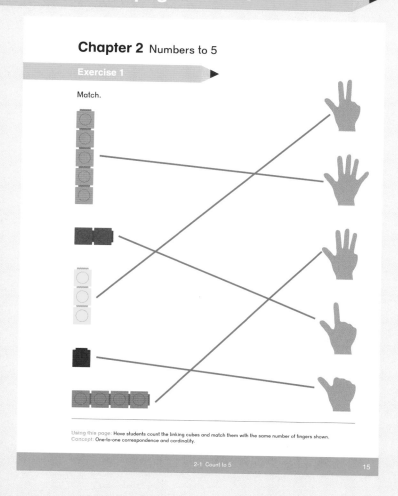

Chapter 2 Numbers to 5

Exercise 1

Match.

Using this page: Have students count the linking cubes and match them with the same number of fingers shown.
Concept: One-to-one correspondence and cardinality.

2-1 Count to 5   15

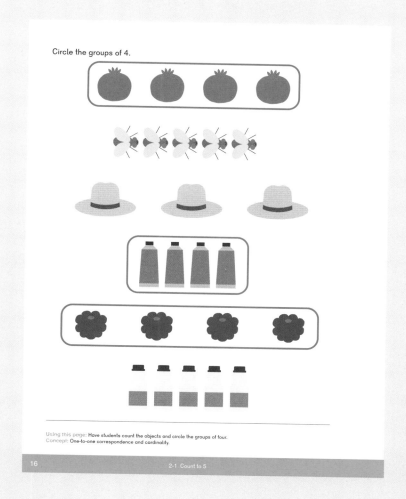

Circle the groups of 4.

Using this page: Have students count the objects and circle the groups of four.
Concept: One-to-one correspondence and cardinality.

16   2-1 Count to 5

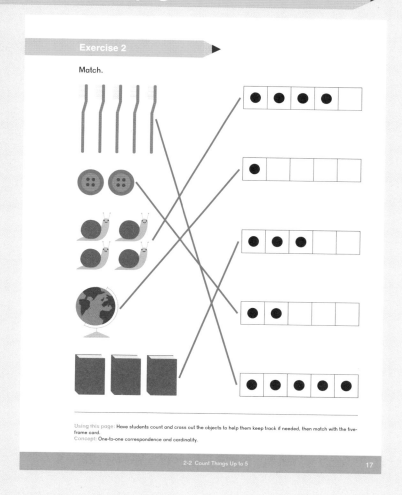

Exercise 2

Match.

Using this page: Have students count and cross out the objects to help them keep track if needed, then match with the five-frame card.
Concept: One-to-one correspondence and cardinality.

2-2 Count Things Up to 5   17

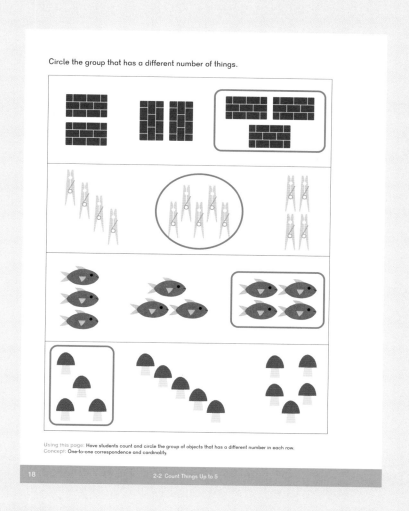

Circle the group that has a different number of things.

Using this page: Have students count and circle the group of objects that has a different number in each row.
Concept: One-to-one correspondence and cardinality.

18   2-2 Count Things Up to 5

Exercise 3

Match.

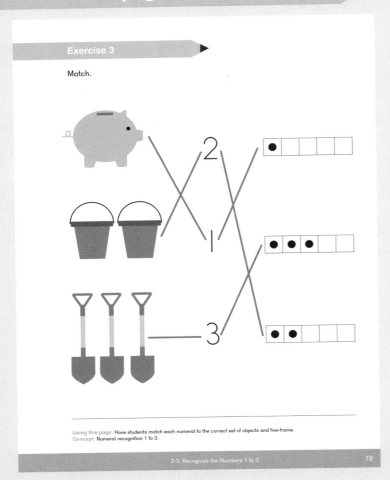

Follow the directions and draw.

Draw 2 balls.

Draw 3 stars.

Draw 1 thing that you like.

Answers will vary.

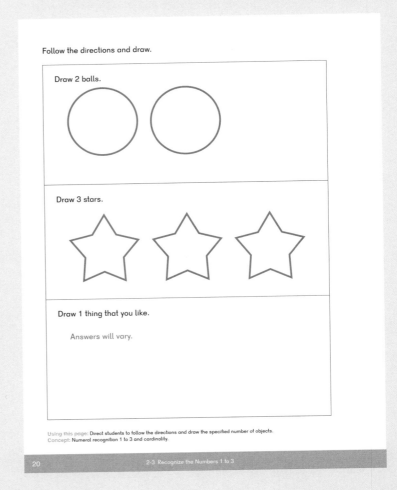

Using this page: Have students match each numeral to the correct set of objects and five-frame.
Concept: Numeral recognition 1 to 3.

Using this page: Direct students to follow the directions and draw the specified number of objects.
Concept: Numeral recognition 1 to 3 and cardinality.

2-3 Recognize the Numbers 1 to 3   19

20   2-3 Recognize the Numbers 1 to 3

Exercise 4

Find and color.
• 4 🐦
• 5 🐤

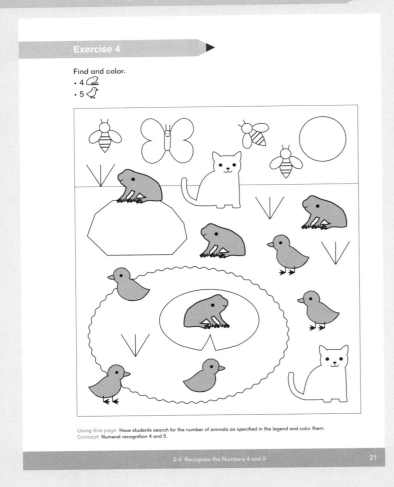

Color according to the Color Key.

Color Key
● 1 ● 2 ● 3 ● 4 ● 5

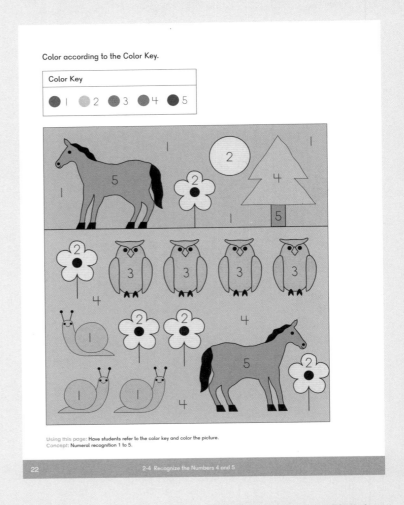

Using this page: Have students search for the number of animals as specified in the legend and color them.
Concept: Numeral recognition 4 and 5.

Using this page: Have students refer to the color key and color the picture.
Concept: Numeral recognition 1 to 5.

2-4 Recognize the Numbers 4 and 5   21

22   2-4 Recognize the Numbers 4 and 5

Exercise 5

Circle the cubes that match the number.

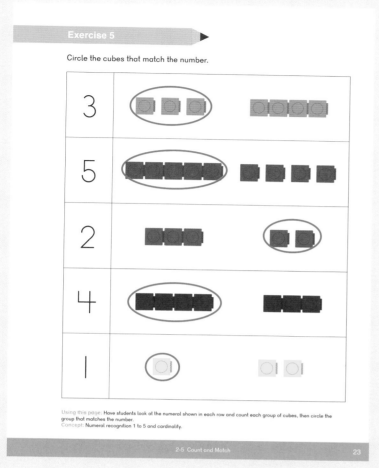

Using this page: Have students look at the numeral shown in each row and count each group of cubes, then circle the group that matches the number.
Concept: Numeral recognition 1 to 5 and cardinality.

2-5 Count and Match   23

Circle the correct number.

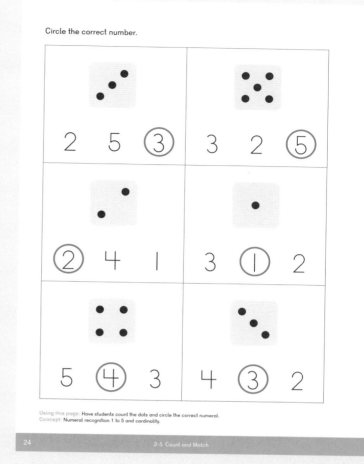

Using this page: Have students count the dots and circle the correct numeral.
Concept: Numeral recognition 1 to 5 and cardinality.

24   2-5 Count and Match

Color the correct number of things.

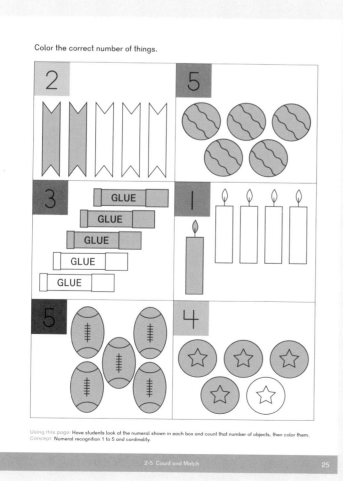

Using this page: Have students look at the numeral shown in each box and count that number of objects, then color them.
Concept: Numeral recognition 1 to 5 and cardinality.

2-5 Count and Match   25

Match.

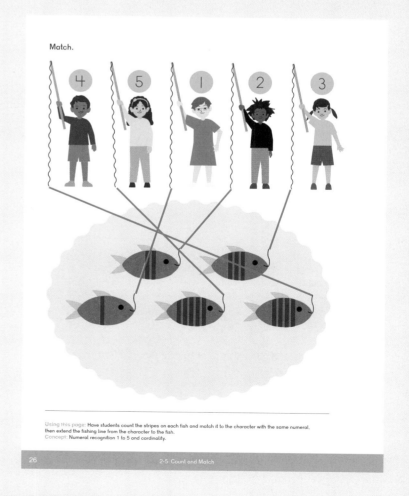

Using this page: Have students count the stripes on each fish and match it to the character with the same numeral, then extend the fishing line from the character to the fish.
Concept: Numeral recognition 1 to 5 and cardinality.

26   2-5 Count and Match

Exercise 6

Trace and write 1.

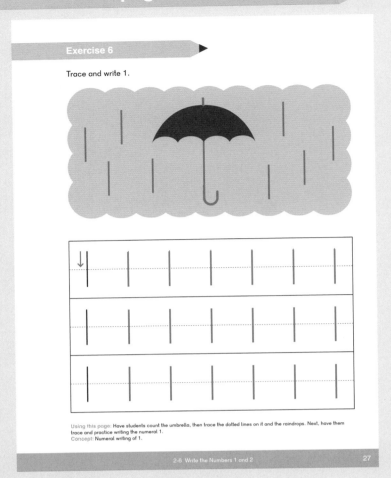

Using this page: Have students count the umbrella, then trace the dotted lines on it and the raindrops. Next, have them trace and practice writing the numeral 1.
Concept: Numeral writing of 1.

2-6 Write the Numbers 1 and 2   27

Trace and write 2.

Using this page: Have students count, then trace the dotted lines around the swans. Next, have them trace and practice writing the numeral 2.
Concept: Numeral writing of 2.

28   2-6 Write the Numbers 1 and 2

Exercise 7

Trace and write 3.

Using this page: Have students count, then trace the dotted lines around the snowmen. Next, have them trace and practice writing the numeral 3.
Concept: Numeral writing of 3.

2-7 Write the Number 3   29

Count and write the number.

Using this page: Have students count the objects and write the numeral in the box.
Concept: Numeral writing of 1 to 3 and cardinality.

30   2-7 Write the Number 3

Exercise 8

Trace and write 4.

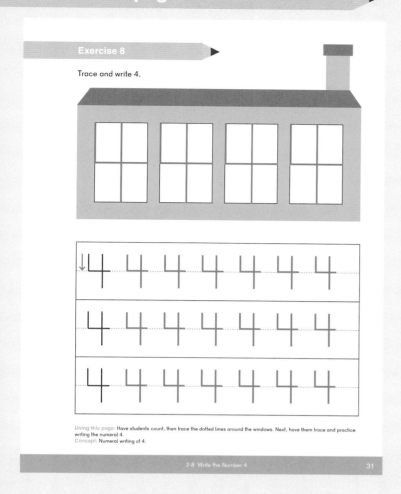

Using this page: Have students count, then trace the dotted lines around the windows. Next, have them trace and practice writing the numeral 4.
Concept: Numeral writing of 4.

2-8 Write the Number 4   31

Count and write the number.

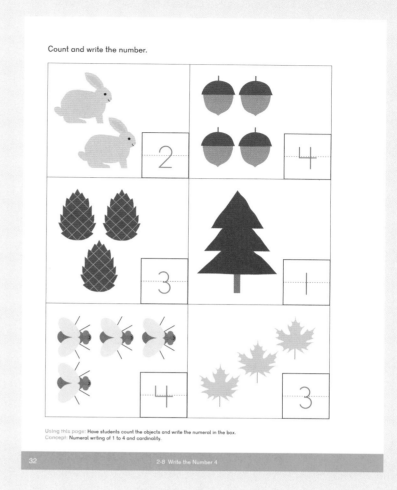

Using this page: Have students count the objects and write the numeral in the box.
Concept: Numeral writing of 1 to 4 and cardinality.

32   2-8 Write the Number 4

Exercise 9

Trace and write 5.

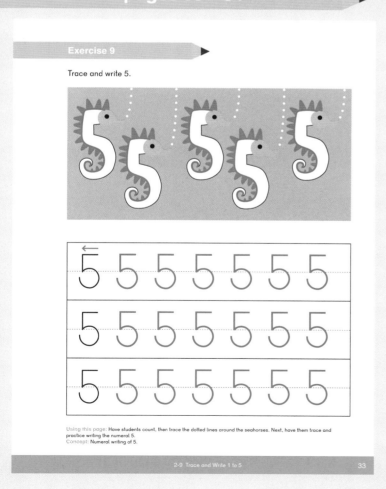

Using this page: Have students count, then trace the dotted lines around the seahorses. Next, have them trace and practice writing the numeral 5.
Concept: Numeral writing of 5.

2-9 Trace and Write 1 to 5   33

Count and write the number.

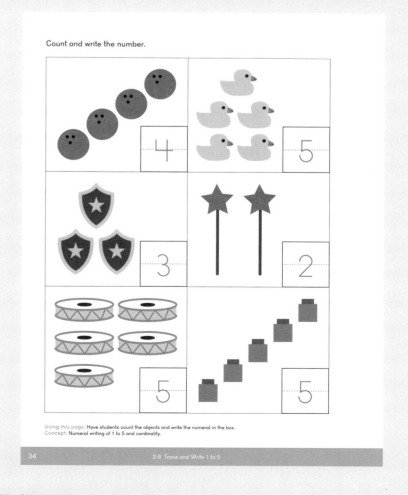

Using this page: Have students count the objects and write the numeral in the box.
Concept: Numeral writing of 1 to 5 and cardinality.

34   2-9 Trace and Write 1 to 5

Exercise 10

Trace and write 0.

Using this page: Have students trace the egg, then determine how many eggs are left after it fell and broke. Next, have students trace and practice writing the numeral 0.
Concept: Numeral writing of 0.

2-10 Zero   35

Count and write the number.

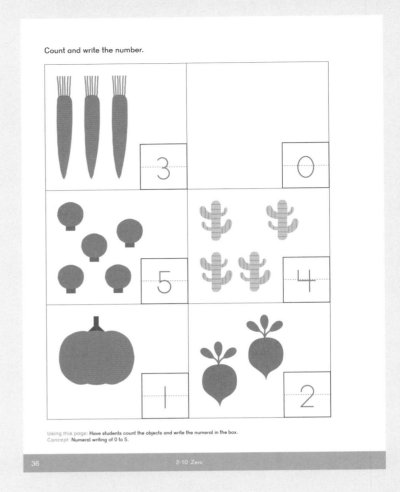

Using this page: Have students count the objects and write the numeral in the box.
Concept: Numeral writing of 0 to 5.

36   2-10 Zero

Exercise 11

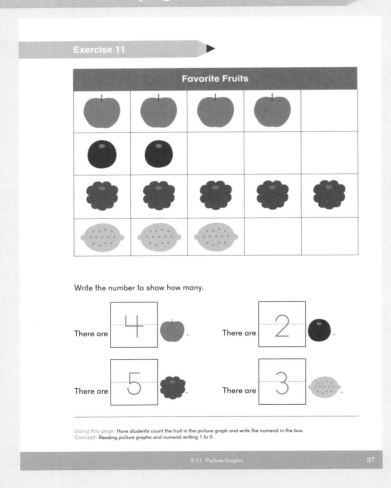

Favorite Fruits

Write the number to show how many.

There are [4] 🍎.   There are [2] ●.

There are [5] 🫐.   There are [3] 🍈.

Using this page: Have students count the fruit in the picture graph and write the numeral in the box.
Concept: Reading picture graphs and numeral writing 1 to 5.

2-11 Picture Graphs   37

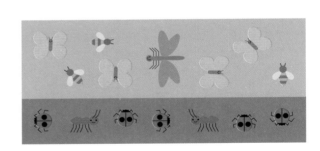

Color the boxes to show how many.

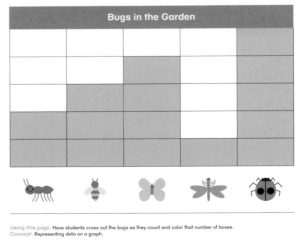

Bugs in the Garden

Using this page: Have students cross out the bugs as they count and color that number of boxes.
Concept: Representing data on a graph.

38   2-11 Picture Graphs

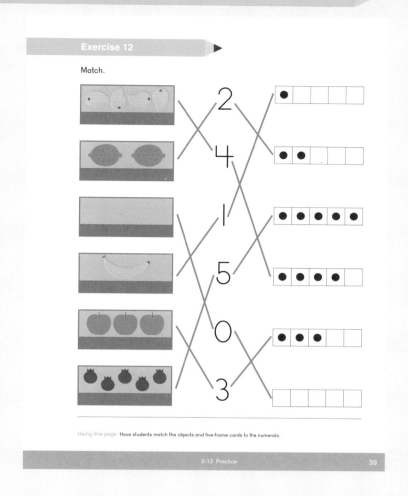

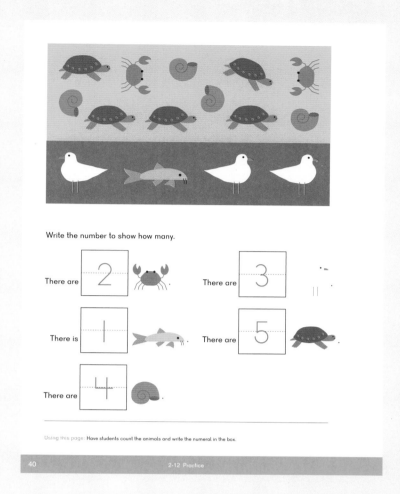

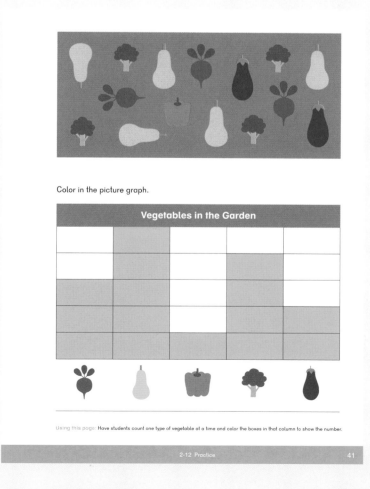

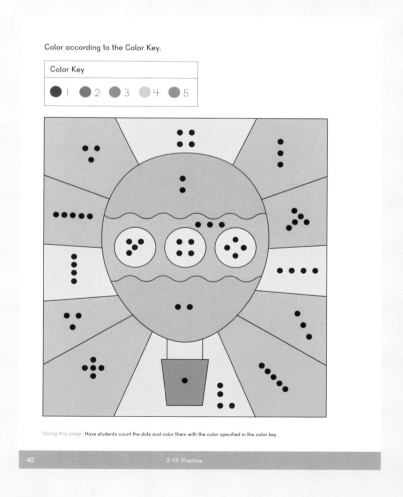

Notes

Suggested number of class periods: 13 – 14

Lesson	Page	Resources	Objectives
Chapter Opener	p. 69	TB: p. 49	
1 Count 1 to 10	p. 70	TB: p. 50	Count 1 to 10 and 10 to 1.
2 Count Up to 7 Things	p. 71	TB: p. 51 WB: p. 43	Count from 1 to 7 with one-to-one correspondence and cardinality.
3 Count Up to 9 Things	p. 73	TB: p. 53 WB: p. 45	Count from 1 to 9 with one-to-one correspondence and cardinality.
4 Counting Up to 10 Things — Part 1	p. 76	TB: p. 56 WB: p. 47	Count from 1 to 10 with one-to-one correspondence and cardinality.
5 Counting Up to 10 Things — Part 2	p. 79	TB: p. 59 WB: p. 49	Count objects up to 10 in different arrangements and represent the number on a ten-frame card.
6 Recognize the Numbers 6 to 10	p. 80	TB: p. 60 WB: p. 53	Recognize the numbers 6 to 10. Match a set of objects to a given number 6 to 10. Represent 6 to 10 on a ten-frame card.
7 Write the Numbers 6 and 7	p. 82	TB: p. 62 WB: p. 55	Write the numerals 6 and 7.
8 Write the Numbers 8, 9, and 10	p. 84	TB: p. 64 WB: p. 57	Write the numerals 8, 9, and 10.
9 Write the Numbers 6 to 10	p. 86	TB: p. 67 WB: p. 61	Write the numerals 6 to 10 to represent the number of objects in a set. Write the numerals 6 to 10 to represent the number represented on a ten-frame card.
10 Count and Write the Numbers 1 to 10	p. 88	TB: p. 69 WB: p. 63	Count with one-to-one correspondence. Practice writing numerals 1 to 10.
11 Ordinal Positions	p. 90	TB: p. 72 WB: p. 67	Recognize ordinal positions first through tenth from various starting points.
12 One More Than	p. 93	TB: p. 75 WB: p. 71	Recognize that numbers in order increase by 1.
13 Practice	p. 95	TB: p. 78 WB: p. 75	Practice concepts from the chapter.
Workbook Solutions	p. 97		

Our system for numeration is a base ten, or decimal system. In a base ten system, there are 10 digits, 0 to 9, and their place in a number is based on powers of 10, e.g. one, tens, hundreds, thousands, etc. So, the number that is one more than nine is written as 10. The 1 in the tens place denotes that it is 1 ten instead of 1 one. Thus, 35 means there are 3 tens and 5 ones.

This system allows for easy computations of large numbers by focusing on only single-digit operations.

Because this system is the basis for all standard algorithms in arithmetic, students are introduced to the concept from the beginning of their math education.

Dimensions Math® Kindergarten focuses on students understanding of the number 10, eventually leading students from understanding 10 as 10 ones to also understanding 10 as 1 ten. From this, they start to understand that the position of the numeral 1 in 10 is important, and will later be able to interpret numbers such as 23 as 2 tens and 3 ones.

In Chapter 3, students will begin using ten-frame cards rather than five-frame cards.

Students will continue learning to write numerals. Special emphasis will be placed on having students notice that even though the number 10 is made up of two digits, it represents one number.

My Book of Numbers

In this chapter, students will have the opportunity to add pages for numbers 6 through 10 to their books.

Parts of this chapter may be a review for some students. Because of the importance of being grounded in 10, plan to play a game or do an activity during every lesson to practice numbers to 10.

Storybooks that go along with content in this chapter can be added as time allows or for additional exploration. Storybook suggestions are listed on the following page.

Lessons on handwriting the numerals 6 through 10 do not have extensions, as it is expected that there will be additional practice on handwriting.

Materials

- Linking cubes
- Beans, beads, bear (or other animal) counters
- Objects to be used as bowling pins
- Small ball
- Pattern blocks

Blackline Masters

- 10 Dots
- Sheep Cards
- My Book of Numbers: Note that once a page for a specific number has been given to students, they will work on that page for multiple lessons. Therefore, new print-outs of My Book of Numbers pages are not required for each lesson.
- Blank Ten-frame
- Ten-frame Cards
- Number Cards
- Number Cards — Large
- Picture Cards
- Blank Graph
- Dot Cards
- Vegetable Cards
- Ordinal Dot Cards
- Ordinal Number Cards
- Number Line 1 to 10

Tactile methods of writing numerals

- Sandpaper: Cut numerals out of sandpaper with a die-cut machine or by hand and glue them to cardstock. Students can either cover the numerals with paper and make numeral rubbings or just trace the numerals with their fingers.
- Whiteboards: Have students write the numerals, then use their index finger to erase.
- Writing in a shallow pan:
  - Filled with sand, rice, or salt
  - Shaving cream writing
  - Gelatin or pudding
- Finger paint writing
- Wikki Stix

Storybooks

- *Ten Black Dots* by Donald Crews
- *Fish Eyes: A Book You Can Count On* by Lois Ehlert
- *Zero Is the Leaves on the Tree* by Betsy Franco
- *My Little Sister Ate One Hare* by Bill Grossman
- *Ten Apples Up On Top!* by Theo LeSieg
- *Count the Ways, Little Brown Bear* by Jonathan London
- *Albert the Muffin-Maker* by Eleanor May
- *Museum 123* by The New York Metropolitan Museum Of Art
- *Mouse Count* by Ellen Stoll Walsh

Letters Home

- Chapter 3 Letter

Notes

Materials

- *Ten Black Dots* by Donald Crews or *Ten Apples Up On Top!* by Theo LeSieg

Explore

Read a number storybook to students. *Ten Black Dots* or *Ten Apples Up On Top!* would be appropriate. Have students count and discuss the pages of the storybook. Share and discuss where we might see objects in groups of tens in our world.

Learn

Discuss page 49. Students should recognize and count 10 fingers, 10 toes, and 10 pins. Some students may see 20 fingers and toes and 20 red lines on the pins. Use this lesson to assess students' prior knowledge of numbers to 10 and one-to-one correspondence when counting objects.

Extend

★ What Could This Be? — Artwork

Materials: Dot markers or 10 cutouts, such as from 10 Dots (BLM) per student

Have students glue 10 cutouts to a page to create a picture, or use 10 Dots (BLM). Students could also use dot markers to create a picture using 10 dots.

Chapter 3

Numbers to 10

49

49

★ What Could This Be? — Story

Materials: Dot markers or 10 cutouts, such as from 10 Dots (BLM) per student

Arrange 10 dots on a piece of copy paper. Make at least three different arrangements of dots and photocopy enough of each arrangement so that each student can choose one. Ask students to embellish the dots on the page to create a picture and/or tell a number story about 10.

Lesson 1 Count 1 to 10

Objective

- Count 1 to 10 and 10 to 1.

Lesson Materials

- Books about counting (see page 67 of this Teacher's Guide)
- Linking cubes, 10 per student

Explore

Invite five students to the front of the class. Identify things on the students that come in pairs. Hands and shoes, for example. Have all students count with you as you point or tap with each count. Ask:

- How many shoes do you see?
- How many hands do you see?
- How many eyes do you see?

Give each student 10 linking cubes. Read a counting book. As numbers appear in the book, have students show that number with their cubes. Rote count from 1 to 10 and then 10 to 1 with the students several times, whispering, shouting, clapping, hopping, stomping, etc.

Learn

Have students discuss Dion's pasture. As they start counting numbers of objects, ask them to show the same number with their cubes. Example:

- I see 10 flowers. (Have all students show 10 cubes.)
- I see 6 fence posts. (Have all students show 6 cubes.)

This is a quick assessment of prior knowledge.

Whole Group Activity

▲ Singing Time

Sing Baa, Baa, Black Sheep.

50

1, 2, 3, 4, 5, 6, 7, 8, 9, 10
Let's start over again.

Objective: Count 1 to 10 by rote and using one-to-one correspondence.

50   3-1 Count 1 to 10

No exercise for this lesson

Extend

★ Missing Cards

Materials: Sheep Cards (BLM) 1 to 10, 1 set per pair of students

Partners use one set of Sheep Cards (BLM). Player 1 removes one card from the set and hands the rest of the cards to Player 2. Player 2 puts the cards in order and figures out the missing card. Players then switch roles. For more of a challenge, have Player 1 remove two cards from the set.

Lesson 2 Count Up to 7 Things

Objective

- Count from 1 to 7 with one-to-one correspondence and cardinality.

Lesson Materials

- Counters or small objects that will fit on a blank ten-frame, 7 per student
- Blank Ten-frames (BLM), 1 per student

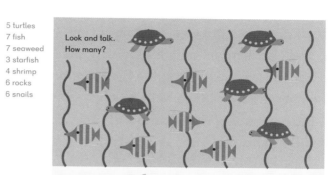

5 turtles
7 fish
7 seaweed
3 starfish
4 shrimp
6 rocks
6 snails

Look and talk.
How many?

Objective: Count up to 7 objects.

3-2 Count Up to 7 Things   51

51

Explore

Provide students with 7 counters each. Ask them to count how many counters there are in all. Note if any students touch or move the same counters more than once and arrive at an incorrect answer. As students count, encourage them to move the counters from the uncounted group to the counted group.

Show students a Blank Ten-frame (BLM). Explain that the ten-frame is similar to the five-frame, but it has 10 boxes. Holding the ten-frame horizontally, ask how many boxes they think are on the top row and on the bottom row.

Provide each student with a Blank Ten-frame (BLM). Ask students to organize their counters on their ten-frame. Ask students to arrange them in different ways and share.

Learn

Have the students look at page 51. Have them count the starfish and show how many starfish there are with counters on their Blank Ten-frame (BLM). Ask them to tell what they did to keep track of their count. Count the starfish together, putting a finger on each as it is counted. Repeat for other objects.

Have the students look at page 52. Have them count the blue, brown, yellow, and purple crayons. Ask them which they need to tap to count and which they already know how many there are just by looking at them. Ask why it is easier to count the crayons than

the objects on the previous page. Continue with the pink and green crayons. Point out the ten-frame cards below the pink and green crayons. Ask how this might help figure out the number.

Whole Group Activities

▲ Listen and Count

Materials: Blank Ten-frame (BLM) for each student, 7 counters or small objects that will fit on a blank ten-frame

Designate a Leader and have her clap 1 to 7 times. Have other students listen and count silently and represent the number of claps heard with counters on their Blank Ten-frame (BLM).

▲ Follow Me

Materials: Number Cards (BLM) 1 to 7

Students take turns drawing a Number Card (BLM) and choosing an action (hopping, clapping, stomping, etc.). For example, the 2 card is drawn and the student chooses to hop. All students hop 2 times. Play continues with the next student drawing a new number card and choosing an action.

Small Group Activities

▲ My Book of Numbers

Materials: My Book of Numbers (BLM) pages 6 and 7, Number Cards (BLM) 6 and 7, Ten-frame Cards (BLM) 6 and 7, Dot Cards (BLM) 6 and 7, Fun Font Number Cards (BLM) 6 and 7, and small objects that can be glued to paper

Give each student the My Book of Numbers (BLM) pages for 6 and 7. Have them decorate each page as in previous lessons. For the 6 page, for example, they will draw six of an item, glue the Number Card (BLM) 6, Ten-frame Card (BLM) 6, Dot Card (BLM) 6, and Fun Font Number Card (BLM) 6 to the page, and glue six small objects down as well. Give students additional pages as needed for any number.

▲ Graph It

Materials: Blank Graph (BLM)

Have students write the names of four of the objects found on page 51 at the bottom of a Blank Graph (BLM), then color in the squares as they count.

Exercise 1 • page 43

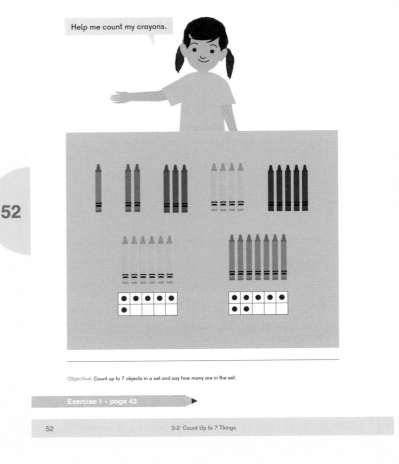

Help me count my crayons.

52

Objective: Count up to 7 objects in a set and say how many are in the set.

Exercise 1 • page 43

52   3-2 Count Up to 7 Things

Extend

★ How Many Can You Make?

Materials: 7 small square colored tiles per student

Using the tiles, students make as many configurations as they can. Sides should touch along their entire length as pictured. Students can record their configurations on graph paper.

Lesson 3 Count Up to 9 Things

Objective

- Count from 1 to 9 with one-to-one correspondence and cardinality.

Lesson Materials

- Picture Cards (BLM) 5 to 9, 1 card per student
- Ten-frame Cards (BLM) 5 to 9, 1 card per student
- Counters or small objects that will fit on a ten-frame, 9 per student
- Blank Ten-frames (BLM), 1 per student

Explore

Pass out Picture Cards (BLM) 5 to 9 and Ten-frame Cards (BLM) 5 to 9 randomly to students, so each student has one card. Have students mingle and find other students with the same number represented either by picture or ten-frame to form five groups. This activity can be repeated multiple times and can be timed to add an element of competition and fun.

Learn

Provide each student with a Blank Ten-frame (BLM) and counters. Have the students look at page 53 and share their observations about the picture, including quantities of objects. Examples:

- 5 children went to the beach.
- There are 8 crabs.

As students share, have all students show the number on their Blank Ten-frame (BLM) with counters.

Have students count Sofia's toys on page 54 and show the number on their Blank Ten-frames (BLM) with counters. Ask students if it's easier to count the balls than the objects on page 53, and to explain their thinking.

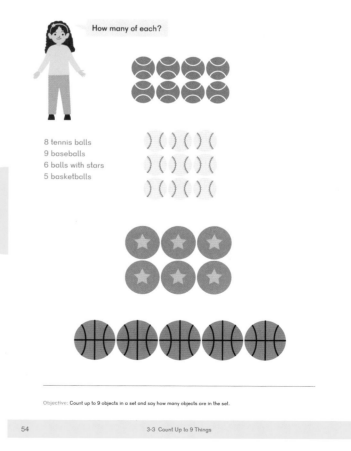

Whole Group Activity

▲ Ten-frame Flash

Materials: Ten-frame Cards (BLM) 1 to 9

Flash Ten-frame Cards (BLM) representing numbers 1 to 9 at the students and have them tell you how many they see.

Small Group Activities

▲ Textbook Page 55

▲ My Book of Numbers

Materials: My Book of Numbers (BLM) pages 8 and 9, Number Cards (BLM) 8 and 9, Ten-frame Cards (BLM) 8 and 9, Dot Cards (BLM) 8 and 9, Fun Font Number Cards (BLM) 8 and 9, and small objects that can be glued to paper

Give each student the My Book of Numbers (BLM) pages for 8 and 9. Have them decorate each page as in the previous lesson.

▲ Order Up — Dot Cards

Materials: Dot Cards (BLM) 1 to 9

Provide pairs of students with one set of Dot Cards (BLM) 1 to 9. Player 1 removes one card from the set and hides it. Player 2 orders the cards from 1 to 9 to determine the missing card. Players switch roles and play continues. For more of a challenge, Player 1 can remove two cards.

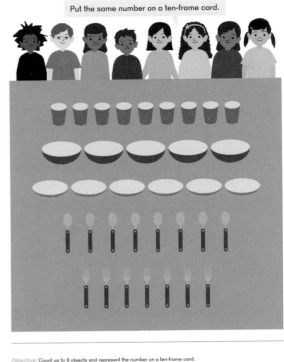

Put the same number on a ten-frame card.

55

Objective: Count up to 9 objects and represent the number on a ten-frame card.

Exercise 2 • page 45

3-3 Count Up to 9 Things   55

▲ Match

Materials: Ten-frame Cards (BLM) 1 to 9 and Dot Cards (BLM) 1 to 9

Students arrange the cards, faceup, in a grid.
Students take turns finding two cards that go together.

★ Memory

Materials: Ten-frame Cards (BLM) 1 to 9 and Dot Cards (BLM) 1 to 9

Students arrange the cards, facedown, in a grid.
Students take turns finding two cards that go together.

Exercise 2 • page 45 ▶

Extend

★ Buzz

Materials: 6-sided die modified with sides 4 – 9, 1 set of Ten-frame Cards (BLM) for each player

Roll the modified die. The number of pips facing up is the target number. Deal the Ten-frame Cards (BLM) facedown equally into stacks for two or three players, and instruct them not to look at their cards.

Players take turns flipping their ten-frame card faceup. If the number on the card matches the target number, players must try to be the first to say, "Buzz." The player who does so wins that card.

After all the cards have been flipped, the player with the most cards wins the round.

Lesson 4 Counting Up to 10 Things — Part 1

Objective

- Count from 1 to 10 with one-to-one correspondence and cardinality.

Lesson Materials

- Linking cubes, 10 per student

Explore

Provide students with 10 linking cubes each. Teach students the song "The Ants Go Marching" (VR). As students sing the song, have them show the same number of cubes as ants marching.

Sing the song a second time and this time have one student at a time bring the same number of cubes as ants marching to the classroom activity space. With each verse, add a row of cubes beneath the previous row. Discuss the patterns in the rows. Point to the fifth row and ask, "If this row has 5, how many are in the next row?" Continue asking through the tenth row.

Learn

Have the students look at page 56. Ask them the question on the page. Have students start at the top row and ask, "How many ants are in that row?" Continue through the tenth row. They should be able to determine the number of ants in each consecutive row without re-counting.

Whole Group Activities

▲ Follow Me

Materials: Number Cards (BLM) 1 to 10

Students take turns drawing a Number Card (BLM) and choosing an action (hopping, clapping, stomping, etc.). For example, the 2 card is drawn and the student chooses to hop. All students hop 2 times. Play continues with the next student drawing a new number card and choosing an action.

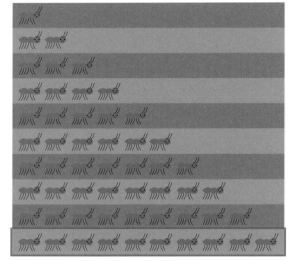

Look and talk.
How many ants in each row?

Objective: Count up to 10 objects in a set and represent the number on a ten-frame card.

56   3-4 Count Up to 10 Things — Part 1

56

Small Group Activities

▲ **Textbook** Page 58

▲ **Bowling**

Materials: Objects to use as bowling pins, such as plastic water bottles, ball to use as a bowling ball

Have students set up "bowling pins" by looking at the **Chapter Opener** on page 49 or page 57. Have students take one turn each to knock down as many pins as they can. They should record the number (or their score) on the whiteboard. After all players have had a turn, the player with the most pins knocked down wins. Players erase their boards and play another round.

▲ **Copy Me**

Materials: Pattern blocks

Using 10 pattern blocks, Player 1 creates a design. Player 2 then tries to duplicate the design. Students switch roles and play continues.

▲ **My Book of Numbers**

Materials: My Book of Numbers (BLM) page for 10, Number Cards (BLM) 10, Ten-frame Cards (BLM) 10, Dot Cards (BLM) 10, Fun Font Number Cards (BLM) 10, and small objects that can be glued to paper

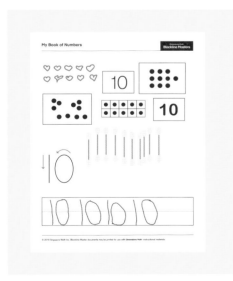

Give each student the My Book of Numbers (BLM) page for 10. Have them decorate each page as in Lesson 2.

Objective: Count up to 10 objects in a set and say how many are in the set.

3-4 Count Up to 10 Things — Part 1   57

▲ Draw and Write

Materials: Number Cards (BLM) 1 to 5, handwriting paper

Students draw from a set of Number Cards (BLM) 1 to 5 and practice writing the number five times on handwriting paper. They then return the card to the set and draw another card to practice writing another number.

▲ Graph It

Materials: Blank Graph (BLM)

Graph It: Have students write the numbers 1 through 10 at the bottom of a Blank Graph (BLM). Students count the ants in each row on page 56 and color the corresponding number of squares above the number.

Exercise 3 • page 47 ▶

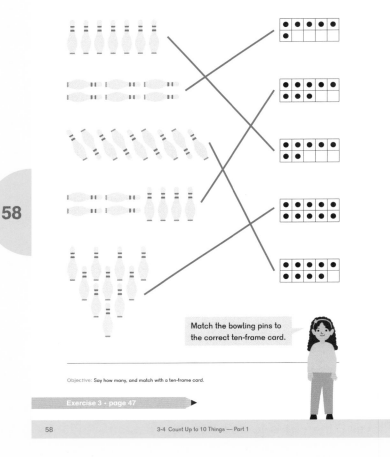

58

Match the bowling pins to the correct ten-frame card.

Objective: Say how many, and match with a ten-frame card.

Exercise 3 • page 47 ▶

58   3-4 Count Up to 10 Things — Part 1

Extend

★ Measuring with Tens

Materials: Linking cubes

Have students use linking cubes to measure the lengths of various items in the classroom by groups of 10.

- How many tens long is the rug?
- How many tens wide is the door?
- How many tens tall is the desk?

Lesson 5 Counting Up to 10 Things — Part 2

Objective

- Count objects up to 10 in different arrangements and represent the number on a ten-frame card.

Lesson Materials

- Bags containing up to 10 counters or other items, 1 per student
- Ten-frame Cards (BLM) 6 to 10, 1 set per student
- Blank Ten-frames (BLM), 1 per student

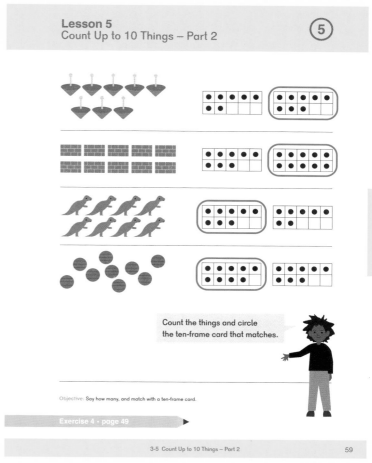

Exercise 4 • page 49

Explore

Fill each bag with between 6 and 10 objects. Give each student one of the bags and a set of Ten-frame Cards (BLM) 6 to 10. Have them find the Ten-frame Card (BLM) that corresponds to the number of objects in their bags. Have students trade bags and repeat.

Whole Group Activity

▲ Listen and Count

Materials: Blank Ten-frame (BLM) for each student, 10 counters or small objects that will fit on a blank ten-frame

Designate a Leader and have him clap 1 to 10 times. Have other students listen and count silently and represent the number of claps heard with counters on their Blank Ten-frame (BLM).

Small Group Activities

▲ Textbook Page 59

▲ How Many Ways?

Materials: Ten-frame Cards (BLM) 6 to 10, Blank Ten-frames (BLM) 6 to 10

Provide students with the cards showing 6 to 10 from sets of Blank Ten-frames (BLM) and Ten-frame Cards (BLM). Students choose a card and color in ten-frames with as many different configurations as they can. Return the card to the set, choose a new card, and repeat.

◄ Exercise 4 • page 49

Lesson 6 Recognize the Numbers 6 to 10

Objectives

- Recognize the numbers 6 to 10.
- Match a set of objects to a given number 6 to 10.
- Represent 6 to 10 on a ten-frame card.

Lesson Materials

- Vegetable Cards (BLM) 6 to 10
- Ten-frame Cards (BLM) 1 to 10
- Number Cards (BLM) 1 to 10
- Blank Ten-frames (BLM)
- Fun Font Number Cards (BLM) 6 to 10

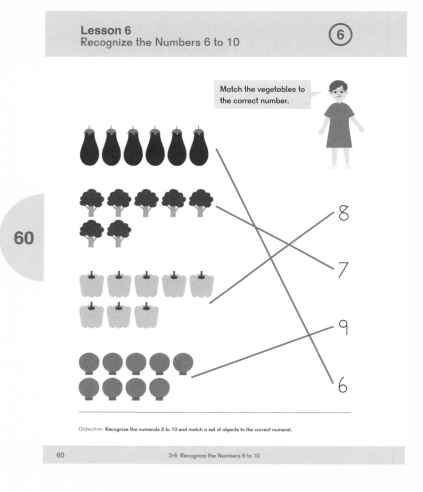

Explore

Pass out Vegetable Cards (BLM) 6 to 10 and Ten-frame Cards (BLM) 6 to 10, so that each student has 1 card.

Have students mingle and find other students with the same number represented either by vegetable or ten-frame to form five groups.

While the students are grouped, show them the Number Card (BLM) that matches the quantity on the group's card. Tell them, "This is the number 6, 7, etc.," as you hold up each card.

Show students Fun Font Number Cards (BLM) 6 to 10 and discuss how the written number is the same or similar. Collect the Vegetable Cards (BLM), then hand out Number Cards (BLM) and repeat the activity.

Learn

Look at page 60 and discuss Emma's vegetables and the numbers. Have students show the number card for the eggplant, then the broccoli, etc.

Whole Group Activity

▲ Follow Me

Materials: Number Cards (BLM) 6 to 10

Students take turns drawing a Number Card (BLM) and choosing an action (hopping, clapping, stomping, etc.). For example, the 2 card is drawn and the student chooses to hop. All students hop 2 times. Play continues with the next student drawing a new number card and choosing an action.

Small Group Activities

▲ **Textbook** Pages 60 – 61

▲ **Count It Out**

Materials: Small objects, Number Cards (BLM) 6 to 10

Provide students with small objects and a set of Number Cards (BLM) 6 to 10. Students choose a card and count out the corresponding number of objects. Play continues until all cards have been used.

▲ **How Many Ways?**

Materials: Ten-frame Cards (BLM) 6 to 10, Dot Cards (BLM) 6 to 10, Blank Ten-frames (BLM)

Provide students with the Blank Ten-frames (BLM) and a deck of Ten-frame Cards (BLM) 6 to 10 and Dot Cards (BLM) 6 to 10. Students choose a Ten-frame Card (BLM) or Dot Card (BLM) and color in Blank Ten-frames (BLM) with as many different configurations as they can. Return the Ten-frame Card (BLM) or Dot Card (BLM) to the deck, choose a new card, and repeat.

▲ **Match**

Materials: Ten-frame Cards (BLM) 1 to 10 and Dot Cards (BLM) 1 to 10

Students arrange the cards, faceup, in a grid. Students take turns finding two cards that go together.

★ **Memory**

Materials: Ten-frame Cards (BLM) 1 to 10 and Dot Cards (BLM) 1 to 10

Students arrange the cards, facedown, in a grid. Students take turns finding two cards that go together.

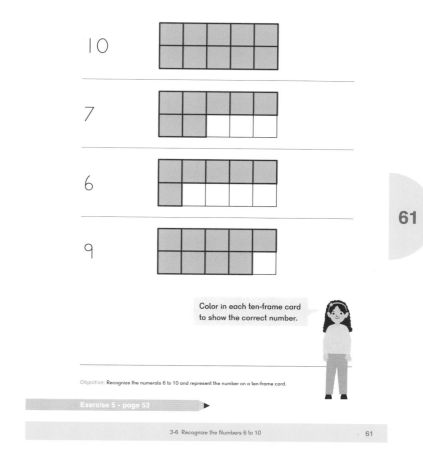

61

Color in each ten-frame card to show the correct number.

Objective: Recognize the numerals 6 to 10 and represent the number on a ten-frame card.

Exercise 5 • page 53

3-6 Recognize the Numbers 6 to 10 · 61

◀ **Exercise 5 • page 53**

Lesson 7 Write the Numbers 6 and 7

Objective

- Write the numerals 6 and 7.

Learn

Students will be writing numerals in this lesson and the next lesson. Students should be at desks or tables to encourage proper posture and hand grip. Beginning on page 62, have students trace the shapes and numerals; then write the numerals, always starting at the top.

Have students trace first with their index finger and then use a pencil to write the numerals.

Students may recite as they write 6:

Start at the top and slide down left.
Then draw a circle to give your 6 some heft.

Look at the triangle on page 63. Students may recite as they write 7:

Across the top, then down the right.
You've got a 7, that's how you write!

Whole Group Activity

▲ Conducting

Materials: Small sticks or straws

Have students use a small stick or straw and become conductors of their number orchestra. Each student forms the numerals in the air.

Small Group Activities

▲ Textbook Pages 62 – 63

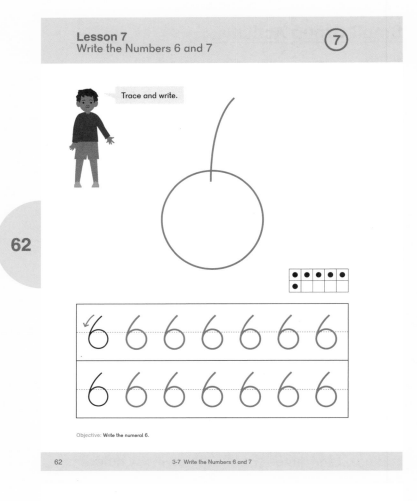

Trace and write.

Objective: Write the numeral 6.

62   3-7 Write the Numbers 6 and 7

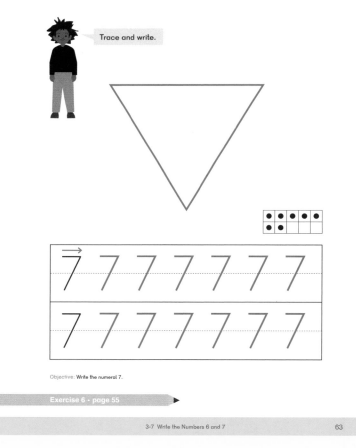

Trace and write.

63

Objective: Write the numeral 7.

Exercise 6 · page 55

3-7 Write the Numbers 6 and 7   63

Centers should include counting small objects to develop fine motor skills and writing with fingers or pencils.

▲ Tactile Writing

Materials: Select from the list below:
- Sandpaper: Cut numerals out of sandpaper with a die-cut machine or by hand and glue them to cardstock. Students can either cover the numerals with paper and make numeral rubbings or just trace the numerals with their fingers.
- Whiteboards: Have students write the numerals, then use their index finger to erase.
- Writing in a shallow pan:
  - Filled with sand, rice, or salt
  - Shaving cream writing
  - Gelatin or pudding
- Finger paint writing
- Wikki Stix

Have students write with their fingers or create numerals out of varied materials.

▲ My Book of Numbers

Materials: My Book of Numbers (BLM) pages 6 and 7

Have each student practice writing the numerals 6 and 7 on the corresponding My Book of Numbers (BLM) pages. Give students additional pages as needed for any number.

▲ Rainbow Writing

Materials: Crayons, markers, or paint

Students practice writing the numerals from the lesson. Each numeral must be written and traced over with at least five different colors so each numeral resembles a rainbow. For students needing assistance, dotted numerals can be traced.

Take it Outside

▲ Walk and Paint the Numbers

Materials: Sidewalk chalk, paintbrushes, cups of water

Use chalk to write very large numerals 6 and 7 as specified in the lesson (about 4 feet tall) outside, each in a different color. Surrounding the large numerals, write more 6s and 7s, but smaller (about 8 inches tall). Students line up single file and walk over the numerals, starting at the top, while saying the rhymes. Provide students with a cup of water and a small paintbrush to trace over the smaller numerals.

◀ **Exercise 6 • page 55**

Lesson 8 Write the Numbers 8, 9, and 10

Objective

- Write the numerals 8, 9, and 10.

Students may recite as they write:

To make an 8, act like a race car.
Drive around four curves but don't go far.
They sit together one on top,
The other on bottom, and then you stop.

Look at the balloons and 9s on page 65. Students may recite as they write:

9 starts with a circle on its left side,
And a nice straight line on the right as a guide.

Look at the flower and 10s on page 66. Students may recite as they write:

The number 10 is like two friends.
And neither of them has sharp bends.
The 1 is a tall straight line from the top.
The 0 is a circle like a round raindrop.

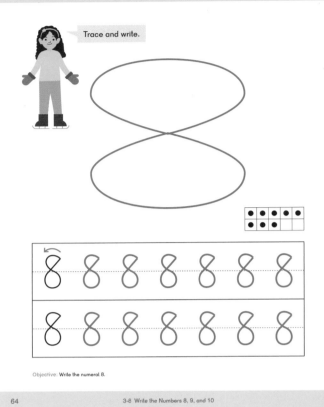

Trace and write.

Objective: Write the numeral 8.

64   3-8 Write the Numbers 8, 9, and 10

Whole Group Activity

▲ Uncover Numbers

Materials: Number Cards — Large (BLM), Blank sheet of paper

Use Number Cards — Large (BLM) and a blank sheet of paper the same size. Choose a number card and cover it with the blank piece of paper. Slowly reveal the number, starting at one of the four corners or the top or bottom. Students shout out the number when they know what it is.

Small Group Activities

▲ Textbook Pages 64 – 66

Centers should include counting small objects to develop fine motor skills and writing with fingers or pencils.

▲ Tactile Writing

Materials: Select from the list below:

- Sandpaper: Cut numerals out of sandpaper with a die-cut machine or by hand and glue them to cardstock. Students can either cover the numerals with paper and make numeral rubbings or just trace the numerals with their fingers.
- Whiteboards: Have students write the numerals, then use their index finger to erase.
- Writing in a shallow pan:
  - Filled with sand, rice, or salt
  - Shaving cream writing
  - Gelatin or pudding
- Finger paint writing
- Wikki Stix

Have students write with their fingers or create numerals out of varied materials.

▲ My Book of Numbers

Materials: My Book of Numbers (BLM) pages 8 – 10

Have each student practice writing the numerals 8, 9, and 10 on the corresponding My Book of Numbers (BLM) pages. Give students additional pages as needed for any number.

Take it Outside

▲ Walk and Paint the Numbers

Materials: Sidewalk chalk, paintbrushes, cups of water

Use chalk to write very large numeral 8 as specified in the lesson (about 4 feet tall) outside, each in a different color. Surrounding the large numerals, write more 8s, but smaller (about 8 inches tall). Students line up single file and walk over the numerals, starting at the top, while saying the rhyme. Provide students with a cup of water and a small paintbrush to trace over the smaller numerals.

Exercise 7 • page 57

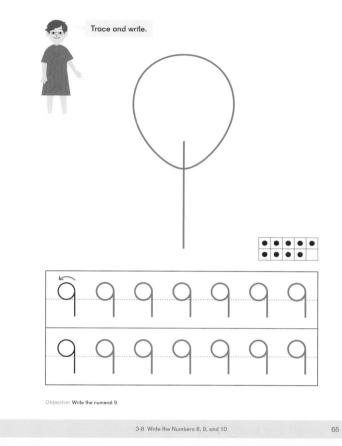

Trace and write.

Objective: Write the numeral 9.

3-6 Write the Numbers 8, 9, and 10   65

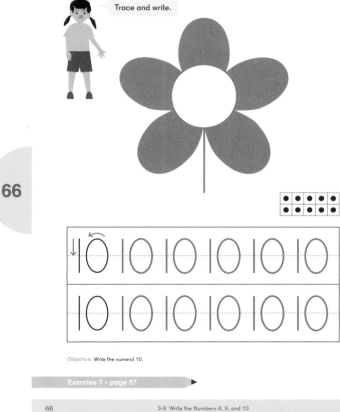

Trace and write.

Objective: Write the numeral 10.

Exercise 7 • page 57

66   3-8 Write the Numbers 8, 9, and 10

Lesson 9 Write the Numbers 6 to 10

Objectives

- Write the numerals 6 to 10 to represent the number of objects in a set.
- Write the numerals 6 to 10 to represent the number represented on a ten-frame card.

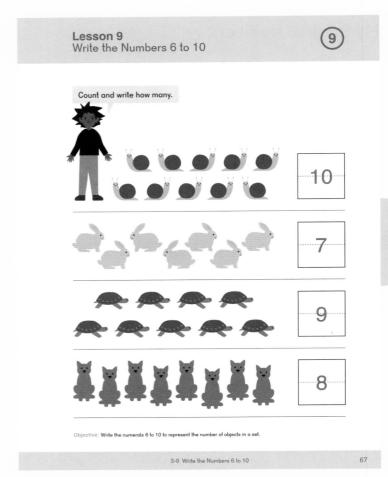

Count and write how many.

10

7

9

8

Objective: Write the numerals 6 to 10 to represent the number of objects in a set.

3-9 Write the Numbers 6 to 10   67

This lesson provides an additional day of practice for counting and writing the numerals 6 to 10.

Whole Group Activities

▲ Listen and Count

Materials: Blank Ten-frame (BLM) for each student, 10 counters or small objects that will fit on a blank ten-frame

Designate a Leader and have her clap 1 to 10 times. Have other students listen and count silently and represent the number of claps heard with counters on their Blank Ten-frame (BLM).

▲ Count the Rabbits

Have students stand in a large circle. While students close their eyes, ask a number of them to open their eyes, go to the middle of the circle, and hop like rabbits. Have students still standing in the circle try to count how many hopping rabbits there are.

Small Group Activities

▲ Textbook Pages 67 – 68

Use activities from prior lessons to continue counting to 10.

▲ Hidden Numbers

Materials: White paper, white crayons, watercolor paints

Have students write a numeral using white crayon on white paper and swap their paper with another student. Students then make a guess on what numeral is written on their swapped paper before painting the paper with watercolor to unveil the mystery number.

Teacher's Guide KA Chapter 3   © 2017 Singapore Math Inc.

▲ Punch Numbers

Materials: Push-pin, golf tee, or hole punch, construction paper or index cards

Have students use a push-pin, golf tee, or hole punch to punch out the outline of numerals 6 to 10 on small pieces of construction paper or index cards.

Exercise 8 • page 61 ▶

68

Extend

★ Number Squeeze

Materials: Number Path 1 to 10 (BLM), counters or game markers

Provide pairs of students with a Number Path 1 to 10 (BLM). Students place a chip (any counter or game marker) on the number path just outside the 1 and the 10. The Chooser chooses a secret number on the number path. The Guessers guess the number and the Chooser says, "My number is less than or greater than" the guessed number and slides the chip from the 1 or the 10 to keep the number squeezed between the two chips. Play continues until the number is either guessed or squeezed out.

For example:

*   The secret number is 2. The Guesser guesses 6 so the chooser says, "My number is less than 6," and slides the chip down from the 10 to the 6 so the secret number remains between the chips.

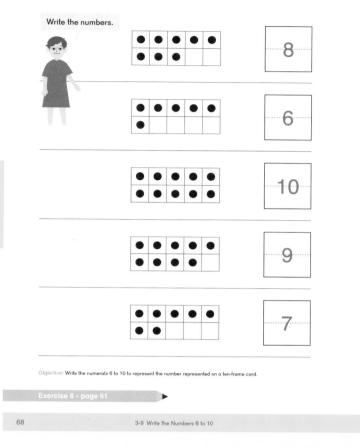

Write the numbers.

8

6

10

9

7

Objective: Write the numerals 6 to 10 to represent the number represented on a ten-frame card.

Exercise 8 • page 61 ▶

68   3-9 Write the Numbers 6 to 10

Lesson 10 Count and Write the Numbers 1 to 10

Objectives

- Count with one-to-one correspondence.
- Practice writing numerals 1 to 10.

Explore

Provide students with tubs of linking cubes. Ask students to grab a small handful of linking cubes, sort them by color, then count and write the numerals on whiteboards or paper. Students return the linking cubes to the tub, grab another small handful, and repeat.

Learn

With partners, have students count and make towers of cubes corresponding to each type of animal on page 69.

Ask students to line the towers up. Ask if they have all the numbers from 1 to 10 or if they're missing some numbers. Have them write the numerals that are missing towers on their whiteboard.

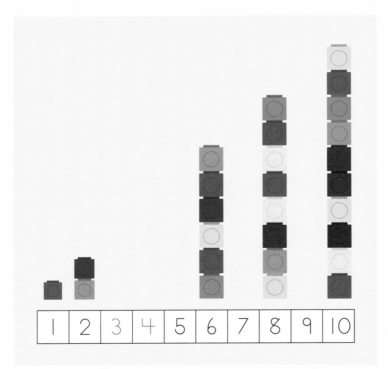

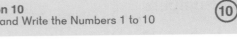

1 horse
2 cows
6 pigs
8 ducks
10 chickens

Look and talk.
How many?

Objective: **Count up to 10 objects.**

3-10 Count and Write the Numbers 1 to 10   69

Whole Group Activity

▲ Flash and Write

Materials: Ten-frame Cards (BLM) 1 to 10

Flash a Ten-frame Card (BLM) and have students write the corresponding numeral on whiteboards. Clear boards and repeat.

Small Group Activities

▲ **Textbook** Pages 70 – 71

▲ **Domino Count**

Materials: Dominoes

Have students choose a domino, count the dots, and write the corresponding numerals on a whiteboard.

Take it Outside

▲ **Circuits**

Students stand in a circle and complete exercises using the numbers 1 to 10 as their count. For example:

- Hop on one foot 1 time
- Do 2 push-ups
- Do 3 jumping jacks
- Do 4 knee bends
- Touch your toes 5 times, etc.

Exercise 9 • page 63

Extend

★ **Hide Your Numbers**

Materials: Art paper, markers or crayons

Have students write the numerals 6 to 10 on a piece of paper, then create a picture around the numerals. After completing their pictures, have them trade with a partner and have the partner find the hidden numbers.

Count and write how many.

3   5   9

2   7

Objective: Write the numerals 1 to 10 to match a set of objects containing that number.

70   3-10 Count and Write the Numbers 1 to 10

Count and write how many.

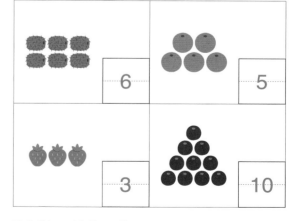

6   5

3   10

Objective: Write a numeral 1 to 10 to represent the number of objects in a set.

Exercise 9 • page 63

3-10 Count and Write the Numbers 1 to 10   71

Lesson 11 Ordinal Positions

Objective

- Recognize ordinal positions first through tenth from various starting points.

Lesson Materials

- *Albert the Muffin-Maker* by Eleanor May or a similar title on ordinal numbers

Explore

Read students *Albert the Muffin-Maker* by Eleanor May or a similar title on ordinal numbers. Discuss the items and language Albert uses in the book: first, second, etc.

Have five students line up in front of the classroom front-to-back with one student designated as the "line leader."

Ask students, "Who is first in line?"

Repeat for student in second place.

Continue naming the positions through fifth place.

Ask students questions with the terms front and back, for example:

- Who is second from the front?
- Who is second from the back?

Have students stand side by side in a line. Ask students similar questions using left and right:

- Who is second from the left?
- Who is second from the right?

Repeat this activity through tenth place.

Learn

Ask students questions about the students in the textbook on page 72. For example, "If Mei is at the front of the line, who is third?"

Lesson 11
Ordinal Positions (11)

Mei is first in line. Who is fourth?

72

Alex
Sofia
Emma
Dion
Mei

Objective: Recognize ordinal positions first through fifth from the front.

72   3-11 Ordinal Positions

Whole Group Activities

▲ Order Up — Students & Cubes

Materials: Linking cubes

Divide the class into equal groups of five or smaller, if necessary. Give each student within the group a different colored linking cube, making sure all the groups have the same colors. Call out a color and a position, starting with "first" and have the student holding the given color within each group line up with their team in the correct position. Examples:

- If you are holding the yellow cube, you are first.
- If you are holding a red cube, line up second.
- If you are holding a blue cube, you are at the back.
- If you are holding a green cube, line up second from the back.

Ask students what their position is. For example, "If green is second from the back, what position is he in from the front?" (third)

Take it Outside

▲ Racing

Have students race in different ways, such as skipping and hopping. After each race, have them announce who came in first through fifth.

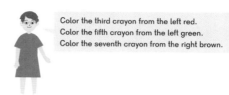

Color the third crayon from the left red.
Color the fifth crayon from the left green.
Color the seventh crayon from the right brown.

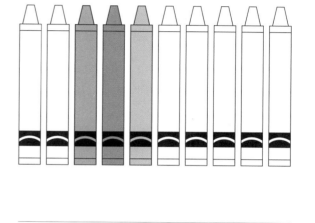

Objective: Identify ordinal positions first through tenth starting from the left or the right.

3-11 Ordinal Positions   73

73

Small Group Activities

▲ **Textbook** Pages 73 – 74

▲ **Gems**

Materials: Counters

Players make a row of counters (or gems) with one red, representing a ruby, and four of a different color, keeping them hidden from their partner. Players take turns asking questions using ordinal words to try to find the ruby. Example:

Player 1: "Is your first counter red?"
Player 2: "No, is your fourth red?"
Player 1: "No, is your second red?"
Player 2: "Yes! You found the ruby!"

▲ Match

Materials: 2 sets of Ordinal Dot Cards (BLM) and Ordinal Number Cards (BLM) 1st to 10th

Students arrange the cards, faceup, in a grid.
Students take turns finding two cards that go together.

★ Memory

Materials: 2 sets of Ordinal Dot Cards (BLM) and Ordinal Number Cards (BLM) 1st to 10th

Students arrange the cards, facedown, in a grid.
Students take turns finding two cards that go together.

Exercise 10 • page 67

Extend

★ A Day in the Life of _____.

Have the students write (or dictate) and illustrate a story, "A Day in the Life of _____" (themselves or a pet). They must use at least one number 1 to 5 and one ordinal position first to fifth in their stories. For example, "The first thing I do in the morning is brush my teeth for 2 minutes. The second thing I do is get dressed."

74

Cross out the third rock from the bottom.
Circle the eighth rock from the top.

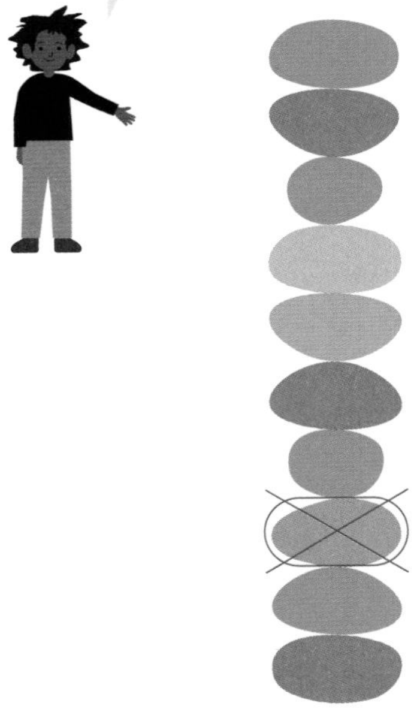

Objective: Identify ordinal positions first through tenth from the top and bottom.

Exercise 10 • page 67

74   3-11 Ordinal Positions

Lesson 12 One More Than

Objective

- Recognize that numbers in order increase by 1.

Lesson Materials

- Linking cubes

Explore

Using linking cubes, have pairs of students work together to make towers of the numbers 1 to 10 and then line them up on their whiteboard in order from 1 to 10. Have them write the numerals under each tower.

Ask students what they notice about the way the towers look. As they are discussing, emphasize that each tower has 1 more cube than the prior tower. Have students say, "1 more than ____ is ____."

Tell them to put a finger on the number 2. Then ask, "What number is 1 more than 2?" Have them move their finger to the 3. Repeat with other numbers.

Learn

Ask students what they notice about the numbers on a number path when considering the numbers from left to right. (The number to the right of any number on the path is one more than that number.)

Have the students look at page 75. Ask them what they notice about the columns that are colored.

- The numbers below each column tell how many squares are colored.
- Each tower (column) has 1 more square colored than the tower to its left.

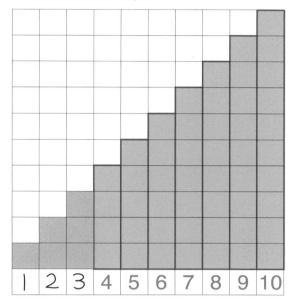

Color the correct number of squares to show one more each time.
Write the number under the squares you colored.
The first three have been done for you.

1 2 3 4 5 6 7 8 9 10

75

Objective: Recognize that numbers in order increase by 1.

3-12 One More Than   75

Small Group Activities

▲ **Textbook** Pages 75 – 77

▲ **Graph It**

Materials: Blank Graph (BLM)

Have students write the names of four colors of triangles found on page 76 at the bottom of a Blank Graph (BLM), then color in the squares as they count.

Exercise 11 • page 71

Extend

★ **Dot Cards**

Materials: Stickers or markers, 10 blank index cards per student

Using stickers or markers on index cards, have students make dot cards with 1 to 10 dots on them. One student flashes a dot card, and the other student identifies the number of stickers on the card; then tells what number is 1 more than that number.

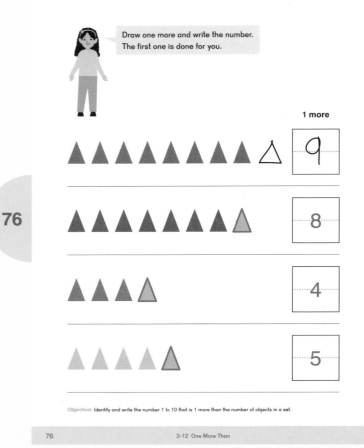

Draw one more and write the number. The first one is done for you.

1 more

9

8

4

5

Objective: Identify and write the number 1 to 10 that is 1 more than the number of objects in a set.

76   3-12 One More Than

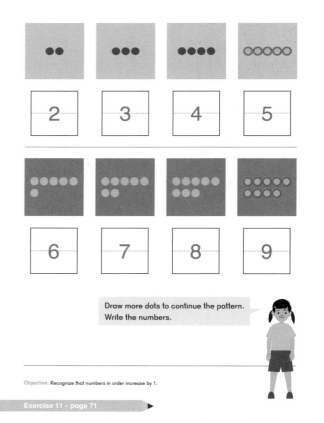

2 3 4 5

6 7 8 9

Draw more dots to continue the pattern. Write the numbers.

Objective: Recognize that numbers in order increase by 1.

Exercise 11 • page 71

3-12 One More Than   77

Lesson 13 Practice

Objective

- Practice concepts from the chapter.

Practice

After students complete the **Practice** in the textbook, have them continue counting and writing numerals from 0 to 10, using **Activities** from the chapter.

Small Group Activity

▲ Number Path Walk

Materials: Sidewalk chalk or paper plates and painter's tape, Ten-frame Cards (BLM) 0 to 10, Dot Cards (BLM) 0 to 10

Create large number paths outside with chalk, or inside with painter's tape or paper plates:

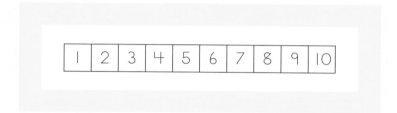

- Have students step on a number that is one more than the number called.
- Show students a Ten-frame Card (BLM) or a Dot Card (BLM) and have them find that number on the number path.

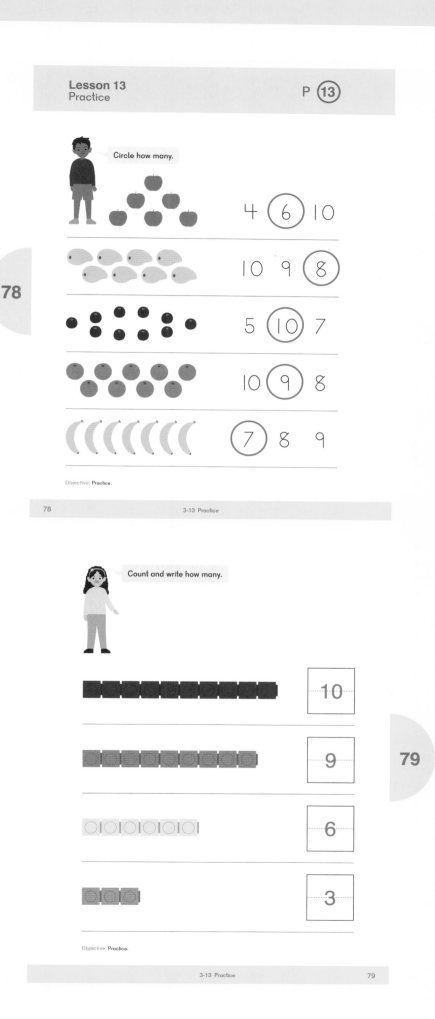

Extend

▲ Match

Materials: Picture Cards (BLM) 0 to 10, Number Cards (BLM) 0 to 10, Ten-frame Cards (BLM) 0 to 10, Number Word Cards (BLM) 0 to 10 for students who are identifying words or reading

Students arrange the cards, faceup, in a grid. Students take turns finding two cards that go together.

★ Memory

Materials: Picture Cards (BLM) 0 to 10, Number Cards (BLM) 0 to 10, Ten-frame Cards (BLM) 0 to 10, Number Word Cards (BLM) 0 to 10 for students who are identifying words or reading

Students arrange the cards, facedown, in a grid. Students take turns finding two cards that go together.

★ Whole Group Activity: Silent Matching

Materials: Picture Cards (BLM) 0 to 10, Number Cards (BLM) 0 to 10, Ten-frame Cards (BLM) 0 to 10, Number Word Cards (BLM) 0 to 10 for students who are identifying words or reading

Pass out a card to each student and have them, without speaking, find other students holding cards that match their own card.

Students can also silently line up in order from 0 to 10 with students holding cards with the same number standing in front of or behind their number match. For example, these cards would all go together:

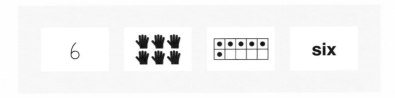

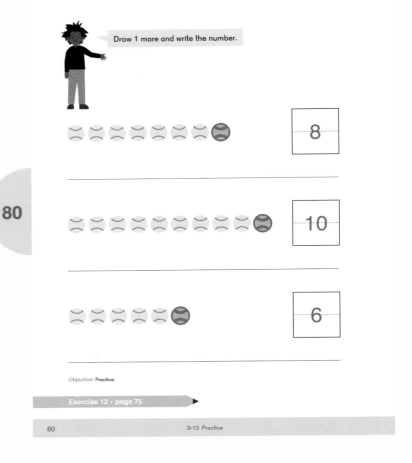

Objective: Practice.

80   3-13 Practice

Looking Ahead

In the **Chapter Opener** for Chapter 4, students explore 3D shapes. To prepare, ask them to bring 3D shapes from home such as boxes and cans (cylinders). They could bring empty and clean packaging material, toys, or other items such as a coin bank shaped like a cylinder.

Chapter 3 Numbers to 10

Exercise 1

Circle the groups of 6.

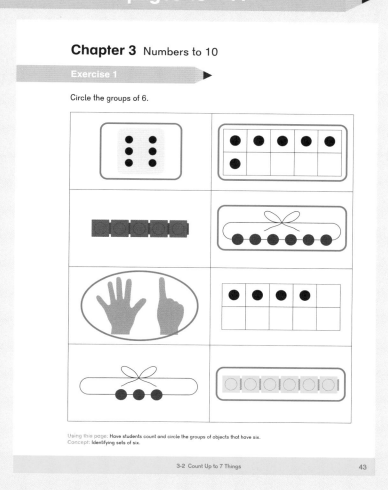

Using this page: Have students count and circle the groups of objects that have six.
Concept: Identifying sets of six.

3-2 Count Up to 7 Things   43

Write an X in the boxes with 7 things.

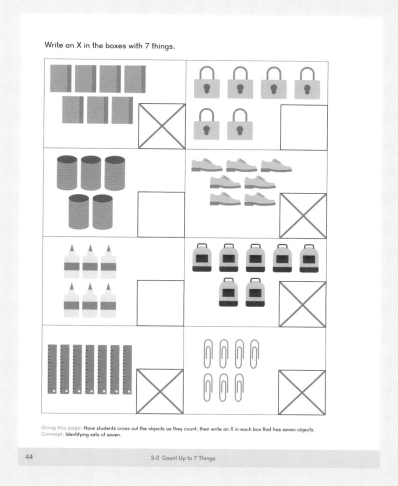

Using this page: Have students cross out the objects as they count, then write an X in each box that has seven objects.
Concept: Identifying sets of seven.

44   3-2 Count Up to 7 Things

Exercise 2

Circle the groups of 8.

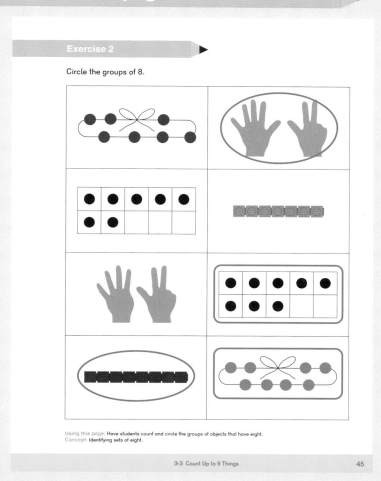

Using this page: Have students count and circle the groups of objects that have eight.
Concept: Identifying sets of eight.

3-3 Count Up to 9 Things   45

Write an X in the boxes with 9 things.

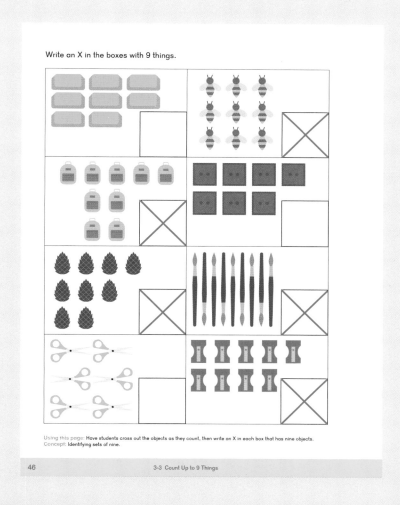

Using this page: Have students cross out the objects as they count, then write an X in each box that has nine objects.
Concept: Identifying sets of nine.

46   3-3 Count Up to 9 Things

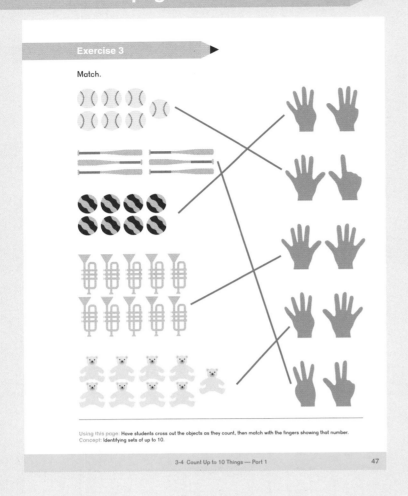

Exercise 3

Match.

Using this page: Have students cross out the objects as they count, then match with the fingers showing that number.
Concept: **Identifying sets of up to 10.**

3-4 Count Up to 10 Things — Part 1   47

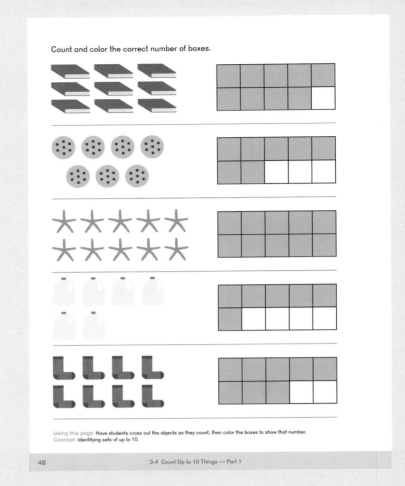

Count and color the correct number of boxes.

Using this page: Have students cross out the objects as they count, then color the boxes to show that number.
Concept: **Identifying sets of up to 10.**

48   3-4 Count Up to 10 Things — Part 1

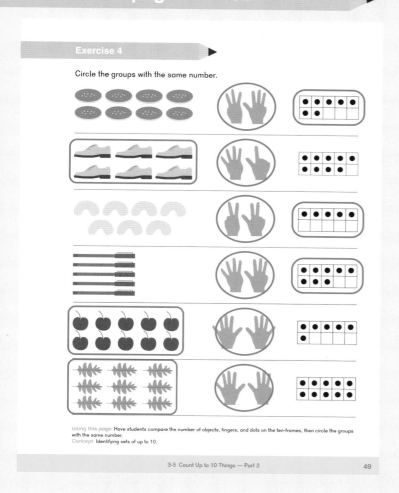

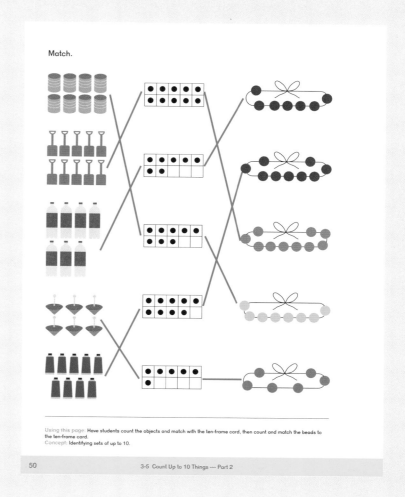

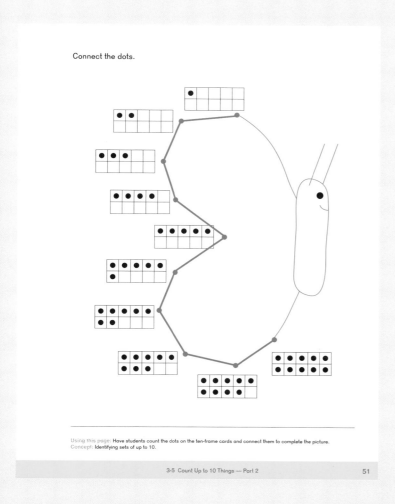

Exercise 5

Circle the correct number.

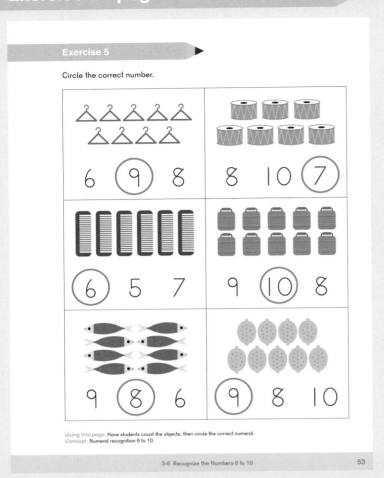

Color according to the Color Key.

Color Key
⬤ 6 ⬤ 7 ⬤ 8 ⬤ 9 ⬤ 10

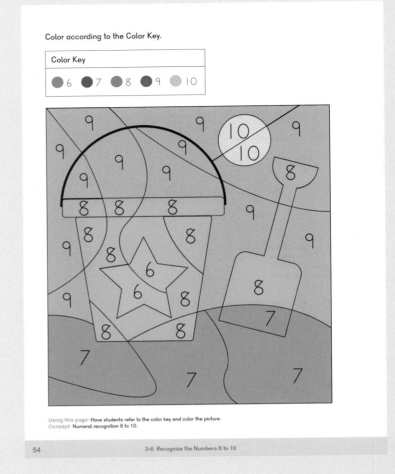

Exercise 6

Trace and write 6.

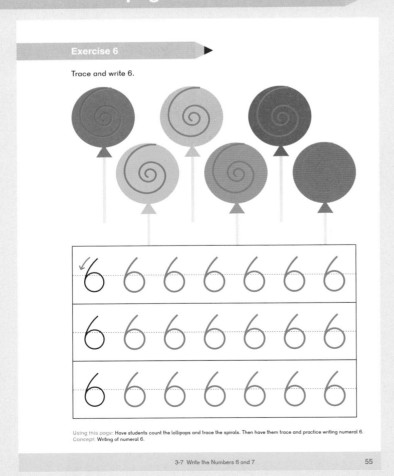

Trace and write 7.

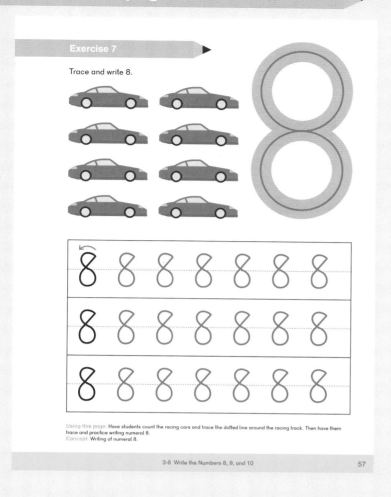

Exercise 7

Trace and write 8.

Using this page: Have students count the racing cars and trace the dotted line around the racing track. Then have them trace and practice writing numeral 8.
Concept: Writing of numeral 8.

3-8 Write the Numbers 8, 9, and 10   57

Trace and write 9.

Using this page: Have students count the dinosaurs and trace the dotted lines around the biggest dinosaur's head. Then have them trace and practice writing numeral 9.
Concept: Writing of numeral 9.

58   3-8 Write the Numbers 8, 9, and 10

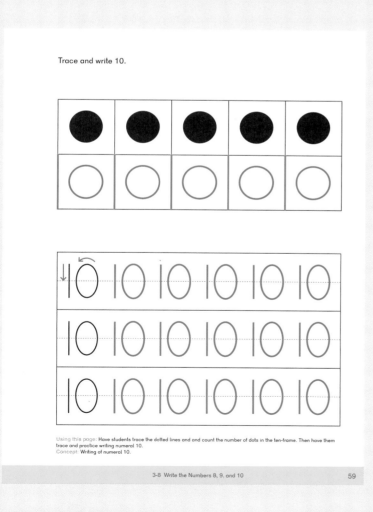

Trace and write 10.

Using this page: Have students trace the dotted lines and and count the number of dots in the ten-frame. Then have them trace and practice writing numeral 10.
Concept: Writing of numeral 10.

3-8 Write the Numbers 8, 9, and 10   59

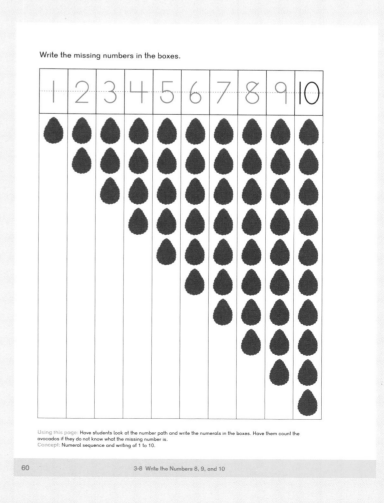

Write the missing numbers in the boxes.

Using this page: Have students look at the number path and write the numerals in the boxes. Have them count the avocados if they do not know what the missing number is.
Concept: Numeral sequence and writing of 1 to 10.

60   3-8 Write the Numbers 8, 9, and 10

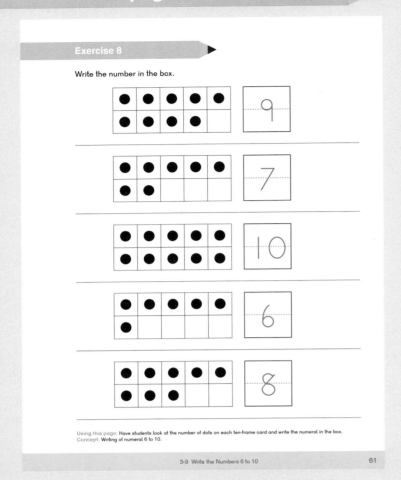

Exercise 8

Write the number in the box.

Using this page: Have students look at the number of dots on each ten-frame card and write the numeral in the box.
Concept: Writing of numeral 6 to 10.

3-9 Write the Numbers 6 to 10   61

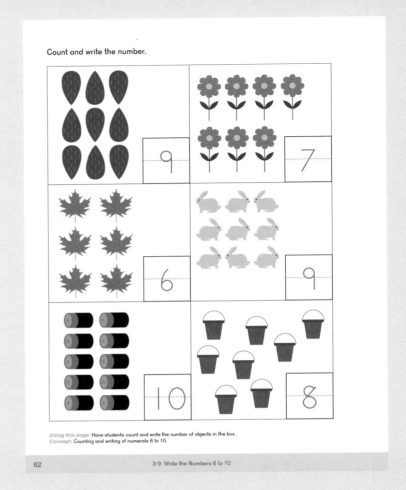

Count and write the number.

Using this page: Have students count and write the number of objects in the box.
Concept: Counting and writing of numerals 6 to 10.

62   3-9 Write the Numbers 6 to 10

Exercise 9

Write the number in the box.

Using this page: Have students count and write the number of fingers in the box.
Concept: Writing of numerals 1 to 10.

3-10 Count and Write the Numbers 1 to 10   63

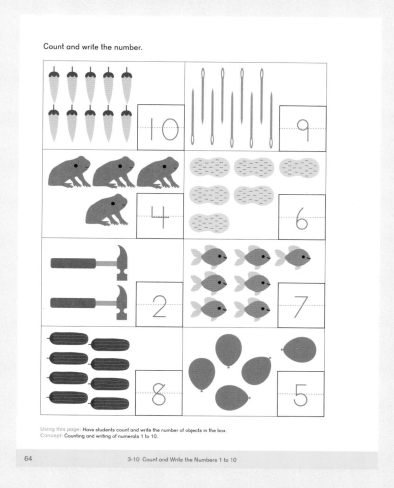

Count and write the number.

Using this page: Have students count and write the number of objects in the box.
Concept: Counting and writing of numerals 1 to 10.

64   3-10 Count and Write the Numbers 1 to 10

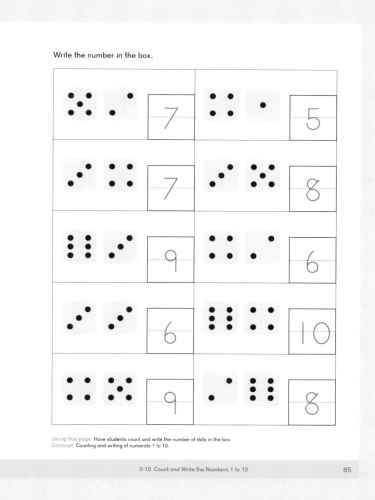

Write the number in the box.

Using this page: Have students count and write the number of dots in the box.
Concept: Counting and writing of numerals 1 to 10.

3-10 Count and Write the Numbers 1 to 10   65

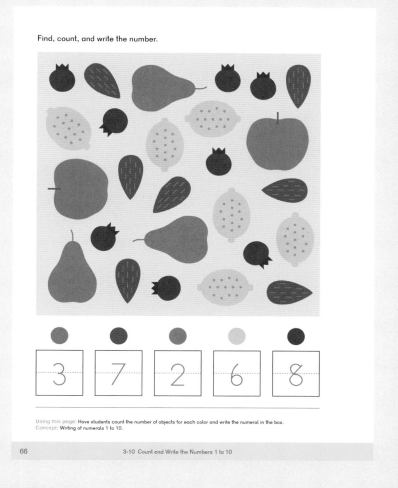

Find, count, and write the number.

Using this page: Have students count the number of objects for each color and write the numeral in the box.
Concept: Writing of numerals 1 to 10.

66   3-10 Count and Write the Numbers 1 to 10

Exercise 10

The first has a ▲.
Count from the front and circle the correct animal.

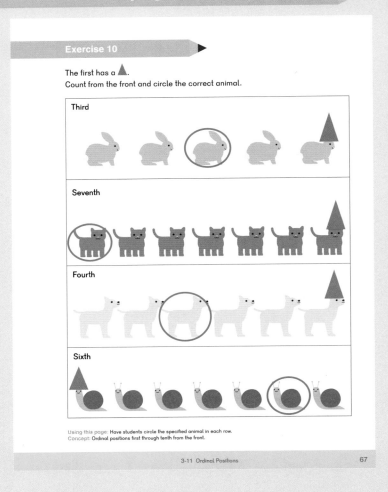

Third

Seventh

Fourth

Sixth

Using this page: Have students circle the specified animal in each row.
Concept: Ordinal positions first through tenth from the front.

3-11 Ordinal Positions   67

Follow the directions and circle.

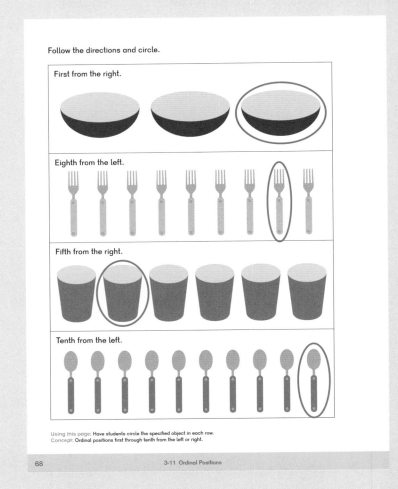

First from the right.

Eighth from the left.

Fifth from the right.

Tenth from the left.

Using this page: Have students circle the specified object in each row.
Concept: Ordinal positions first through tenth from the left or right.

68   3-11 Ordinal Positions

Count from the top and match.

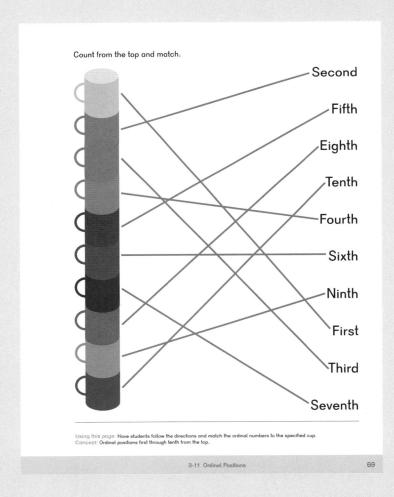

Second

Fifth

Eighth

Tenth

Fourth

Sixth

Ninth

First

Third

Seventh

Using this page: Have students follow the directions and match the ordinal numbers to the specified cup.
Concept: Ordinal positions first through tenth from the top.

3-11 Ordinal Positions   69

Count from the bottom and color according to the Color Key.

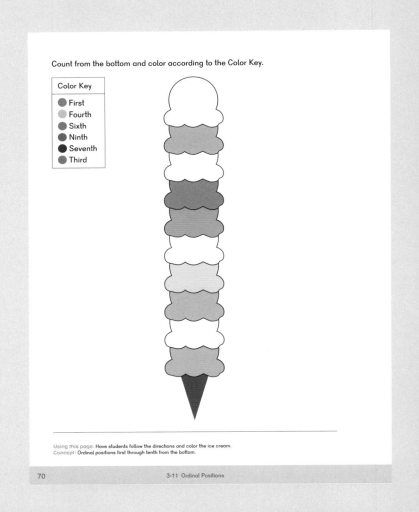

Color Key
● First
● Fourth
● Sixth
● Ninth
● Seventh
● Third

Using this page: Have students follow the directions and color the ice cream.
Concept: Ordinal positions first through tenth from the bottom.

70   3-11 Ordinal Positions

Exercise 11

Circle the group that has 1 more in each box.

Using this page: Have students compare the two groups of objects in each box and circle the group that has one more than the other group.
Concept: Each successive number name refers to a quantity that is one more than the previous quantity.

3-12 One More Than   71

Count and write the numbers.
Circle the number that is 1 more.

	5
	6
	8
	7
	4
	3
	9
	10

Using this page: Have students count each group of objects and write the numeral in the box. Then have them circle the number that is one more than the other.
Concept: Each successive number name refers to a quantity that is one more than the previous quantity.

72   3-12 One More Than

Circle the group that has 1 more than the number.

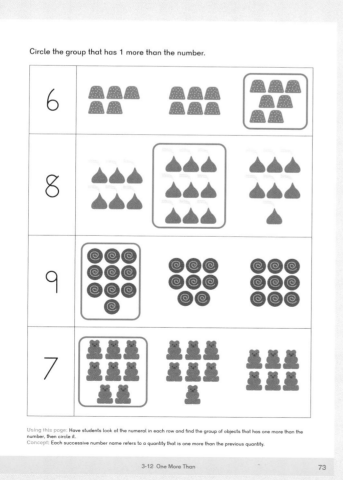

Using this page: Have students look at the numeral in each row and find the group of objects that has one more than the number, then circle it.
Concept: Each successive number name refers to a quantity that is one more than the previous quantity.

3-12 One More Than   73

Circle the number that comes next.

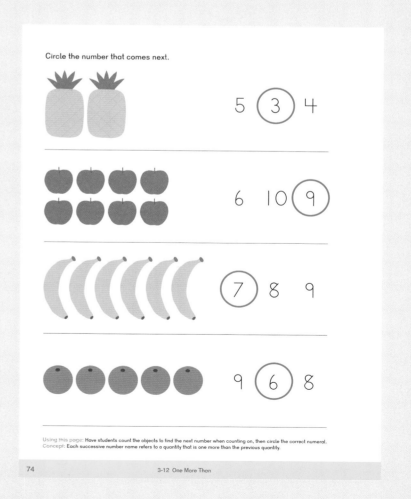

Using this page: Have students count the objects to find the next number when counting on, then circle the correct numeral.
Concept: Each successive number name refers to a quantity that is one more than the previous quantity.

74   3-12 One More Than

Exercise 12

Connect the dots.

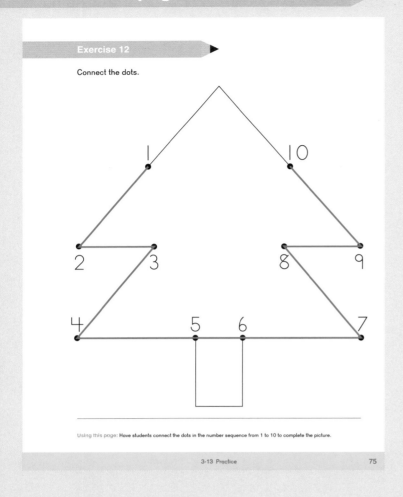

Using this page: Have students connect the dots in the number sequence from 1 to 10 to complete the picture.

3-13 Practice 75

Write the number in the box.

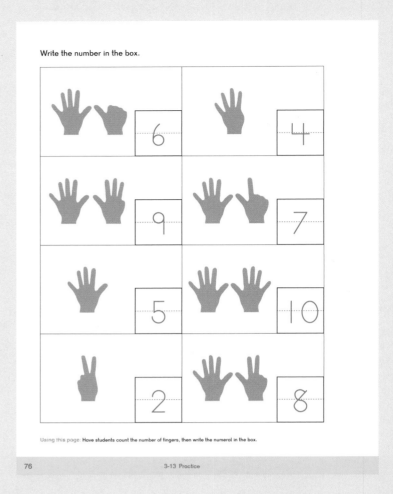

Using this page: Have students count the number of fingers, then write the numeral in the box.

76 3-13 Practice

Match the number with the group that has 1 more.

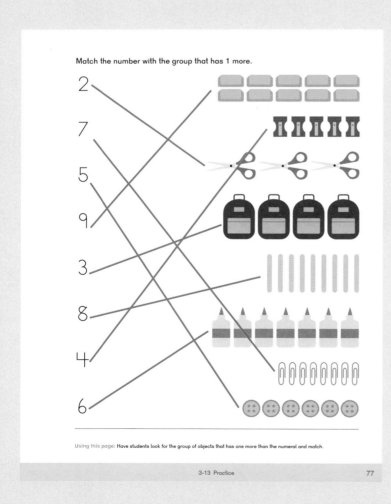

Using this page: Have students look for the group of objects that has one more than the numeral and match.

3-13 Practice 77

Count, write, and circle the group that comes next.

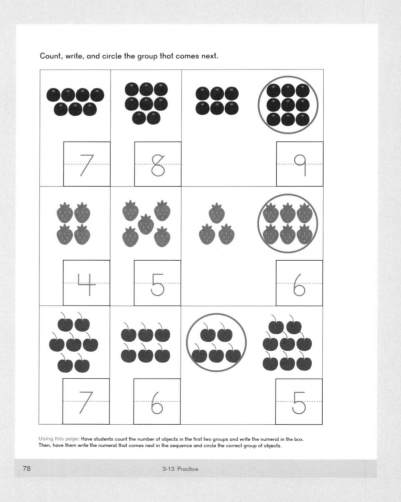

Using this page: Have students count the number of objects in the first two groups and write the numeral in the box.
Then, have them write the numeral that comes next in the sequence and circle the correct group of objects.

78 3-13 Practice

Follow the directions and color the fence.
Color the fifth from the left ●.
Color the seventh from the right ●.
Color the ninth from the left ○.

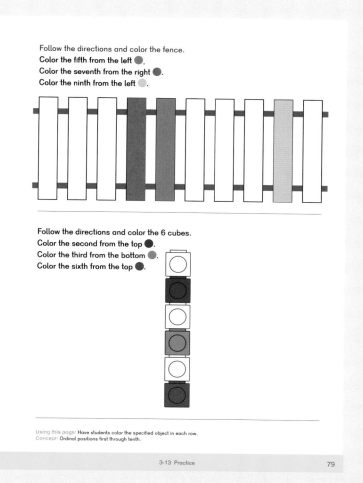

Follow the directions and color the 6 cubes.
Color the second from the top ●.
Color the third from the bottom ●.
Color the sixth from the top ●.

Find, count, and write the number.

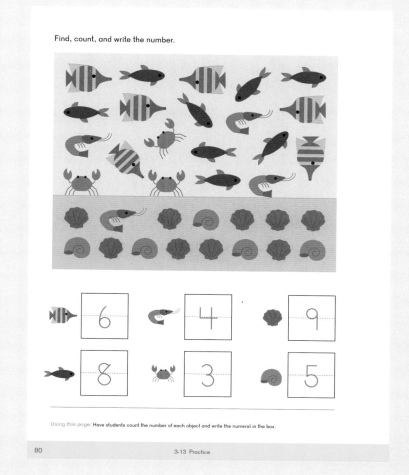

Using this page: Have students color the specified object in each row.
Concept: Ordinal positions first through tenth.

Using this page: Have students count the number of each object and write the numeral in the box.

Notes

Suggested number of class periods: 12 – 13

Lesson		Page	Resources		Objectives
	Chapter Opener	p. 113	TB:	p. 81	
1	Curved or Flat	p. 114	TB:	p. 82	Identify curved and flat surfaces. Build with solids.
2	Solid Shapes	p. 116	TB: WB:	p. 83 p. 81	Identify cubes, cylinders, spheres, and cones.
3	Closed Shapes	p. 118	TB: WB:	p. 85 p. 83	Identify closed shapes.
4	Rectangles	p. 119	TB: WB:	p. 86 p. 85	Identify rectangles.
5	Squares	p. 121	TB: WB:	p. 88 p. 87	Identify squares.
6	Circles and Triangles	p. 123	TB: WB:	p. 89 p. 89	Identify circles and triangles.
7	Where is It?	p. 126	TB: WB:	p. 92 p. 93	Use position words to describe the location of an object or shape.
8	Hexagons	p. 128	TB: WB:	p. 94 p. 95	Identify and name hexagons. Copy a pattern.
9	Sizes and Shapes	p. 130	TB: WB:	p. 96 p. 97	Identify shapes and solids according to size. Continue a pattern.
10	Combine Shapes	p. 132	TB: WB:	p. 98 p. 99	Make new shapes by combining basic shapes.
11	Graphs	p. 134	TB: WB:	p.100 p. 101	Represent data in a graph.
12	Practice	p. 136	TB: WB:	p. 102 p. 103	Practice concepts from the chapter.
	Workbook Solutions	p. 138			

In this chapter, students will recognize and investigate basic two-dimensional (2D) and three-dimensional (3D) shapes, and identify these shapes in the world. The focus will be on beginning to recognize attributes of shapes, such as boundaries, sides, corners, size, orientation, and position relative to other objects. Students will not be expected to learn formal properties and definitions of shapes, nor to name parallel or perpendicular sides or right angles.

Students will learn to describe shapes by their attributes formally in later grades, at which time they will learn that many of the objects from the environment they are identifying as common shapes are approximations of those shapes, as the formal definitions will not apply. For example, at this time they can identify a cell phone as having the shape of a rectangle, but it is not a true rectangle because it has rounded corners.

One instance where formal properties are used is in the explanation of squares as special rectangles. Children at this stage typically recognize shapes by their looks. Squares and rectangles look different to them since a square always has equal sides and so always looks the same, except bigger or smaller. Rectangles can, however, be long and skinny, or short and wide. Later, students will learn that the formal geometric definition of a rectangle applies to squares as well, and that there is a hierarchy in the classification of shapes.

For this reason, students in **Dimensions Math®** **Kindergarten** are told directly that squares are special rectangles with four sides all the same.

Lessons in the latter part of this chapter cover patterns and include hexagons as an example of a common shape with more than four sides.

Many of the lessons are hands-on and will incorporate activities for groups within the lesson. As such, there are fewer suggestions for <u>**Whole Group Activities**</u> and <u>**Small Group Activities**</u> in this chapter. Invite students to bring objects from home to add to a classroom "shape museum" for students to explore throughout the chapter. Additionally, activities that practice and review numbers to 10 are also included.

Storybooks that go along with content in this chapter can be added as time allows or for additional exploration. Storybook suggestions are listed on the following page.

Note: While not introduced in this level or in the chapter, the term used for non-cube "box" shapes is either "cuboid" or "rectangular prism."

Materials

- Three-dimensional shapes: cuboids (rectangular prisms), cylinders, cubes, and cones
- Objects from home that look like one of the 3D shapes: cuboids, pyramids, cylinders, cubes, and cones, such as snack boxes, chip cylinders, cookie tins, etc.
- Paper bag or sock
- Dominoes
- Pipe cleaners or chenille stems
- Geoboards
- Rubber bands
- Index cards
- Linking cubes
- Pre-cut paper or foam shapes: rectangles, squares, circles, and triangles
- Attribute blocks
- Pattern blocks
- Play dough or salt dough
- Tangram sets
- Poster paper and markers
- Painter's tape

Blackline Masters

- Die
- Geoboard Dot Paper
- Number Cards
- Tangram Pieces
- Blank Graph
- Shapes & Sizes Picture Cards

Storybooks

- *The Shape Of Things* by Dayle Ann Dodds
- *Cubes, Cones, Cylinders, & Spheres* by Tana Hoban
- *Shapes, Shapes, Shapes* by Tana Hoban
- *Three Pigs, One Wolf, and Seven Magic Shapes* by Grace Maccarone
- *Shape by Shape* by Suse MacDonald
- *I Spy Shapes in Art* by Lucy Micklethwait
- *Skippyjon Jones Shape Up* by Judy Schachner
- *Mouse Shapes* by Ellen Stoll Walsh
- *The Shape of Me and Other Stuff* by Dr. Seuss
- *Goldilocks and the Three Bears*

Letters Home

- Chapter 4 Letter

Notes

Lesson Materials

- Boxes and cylinders for students to use for building (shoe boxes, tissue boxes, paper towel rolls, cans, etc.)

Shapes and Solids

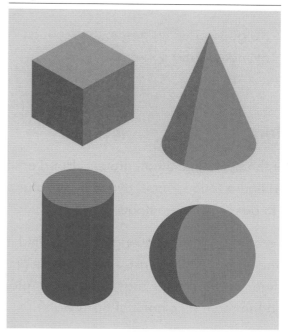

81

Explore

Collect the 3D shapes brought from home, such as boxes and cans, and have students use them, or objects found in the classroom, to build a structure.

Discuss with students the structures they built and the attributes of the 3D shapes. Examples:

- Which ones roll? Which ones slide?
- Which ones look like boxes?
- Which ones have circles on them?

The purpose of this activity is to informally explore attributes of solids, not to name them, in preparation for **Lesson 1: Curved or Flat** and **Lesson 2: Solid Shapes**, where they will identify and name cubes, spheres, cones, and cylinders.

Learn

Look at and discuss the 3D shapes on page 81. Ask, "How are these the same or different?," "Which is most like the object you found and why?"

In this **Chapter Opener** and **Lesson 1: Curved or Flat**, students may refer to all surfaces, both flat and curved faces of a 3D shape as "surfaces." They do not need to learn the formal definition of a "face."

Extend

★ **What Could This Be? — Art**

Materials: 3D shapes

Give each student a 3D shape and ask, "What could this be?" Invite students to share orally and then draw a picture of their idea.

▲ **What Could This Be? — Construction**

Materials: 3D shapes, blocks

Challenge students to build structures with classroom blocks using any given combination of 3D shapes. Can they build a structure using 3 cones?

Lesson 1 Curved or Flat

Objectives

- Identify curved and flat surfaces.
- Build with solids.

Lesson Materials

- 3D geometric shapes: cubes, cones, cylinders, and spheres

Explore

Divide the class into small groups. Hand each group a solid figure (cube, cone, cylinder, or sphere). Ask them to describe their shape.

Have groups go on a scavenger hunt to find items from the classroom that are similar to the group's solid figure. Some items may need to be "planted" around the room (like conical party hats).

Discuss what students notice about the objects. Focus attention on the surfaces of the items and introduce the terms "flat" and "curved." Have students say how many flat or curved surfaces there are on each shape.

Learn

Have students identify the flat and curved surfaces on Emma's shapes. Ask students what we might call the place where the surfaces meet. Suggest that the term "edge" may be less confusing than "side." Discuss the edges on the shapes in the textbook. Students will name the shapes in the next lesson.

Lesson 1
Curved or Flat
①

Look and talk.
Are the surfaces curved or flat?

82

Objective: Identify curved and flat surfaces and build with some solids.

82   4-1 Curved or Flat

Whole Group Activity

▲ Shape Riddles

Teacher gives a series of clues describing a solid figure, one at a time, pausing to take a guess from students between each clue. For example, the teacher may give the following series of clues:

- It has a curved surface.
- It also has one flat surface.
- It has a point.
- I might wear it at a party.

Small Group Activities

▲ Roll or Slide?

Materials: 3D geometric shapes: cubes, cones, cylinders, and spheres

Invite students to build ramps and experiment with 3D shapes. Ask students, "If a flat surface is in contact with the ramp, does it roll or slide?" and, "Which type of surface will allow the shape to roll? To slide?" Students can sort 3D shapes into roll or slide groups.

▲ Domino Sort

Materials: Dominoes up to double 5

Provide students dominoes. Have students sort dominoes by the number of pips. This activity is a review of numbers to 10.

Extend

★ What Could This Be? — Extension

Materials: Die (BLM)

Have students decorate, then cut, fold, and glue the Die (BLM) to use for games. Ideas for decorating the die:

- Numerals 0 to 5
- Pips for 0 to 5
- Pictures for 0 to 5
- Shapes or colors

No exercise for this lesson ▶

Objective

- Identify cubes, cylinders, spheres, and cones.

Lesson Materials

- 3D geometric shapes: cubes, cones, cylinders, and spheres
- Paper bag or sock

Explore

Show examples of the four basic solids. Ask students what they notice, and highlight curved and flat surfaces and edges. For example, a cube:

- Has flat surfaces.
- Has 6 surfaces that are all the same.
- Has edges and corners.

Contrast the cube with the sphere and ask students what they notice:

- Its surface is curved.
- It doesn't have any edges or corners at all.

Have students discuss the cylinder and cone. Ask students how many surfaces the sphere and cone have.

Using objects found in the classroom or brought to school for the **Chapter Opener**, have students sort solids by shape. **Note**: Students don't need to be able to use the names of the shapes. This is an introduction, and hearing the correct terms along with identifying common attributes is sufficient.

Learn

Look at page 83. Ask students what the objects in the first column are. Ask what shape in the row most closely resembles the shape of each object. Remind students that the shapes will be similar, not the same. For example, the pot has handles, but most of the pot is similar to the cylinder.

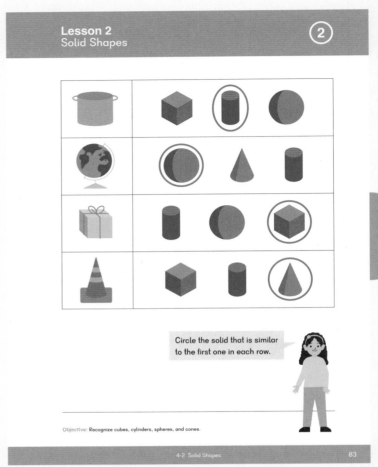

Lesson 2
Solid Shapes   ②

Circle the solid that is similar to the first one in each row.

Objective: Recognize cubes, cylinders, spheres, and cones.

4-2 Solid Shapes   83

Whole Group Activity

▲ Three-dimensional Riddles

Materials: 3D geometric shapes: cubes, cones, cylinders, and spheres

Give students 3 clues about an object. Students have to guess the 3D shape of the object. For example:

1. I am made of paper.
2. I am colorful.
3. You wear me on your head at a party.
   What 3D shape could I be?

Students respond, "A cone."

Alternatively, if students are not yet able to identify the shapes by name, students could identify the 3D shape from a group of shapes.

Small Group Activities

▲ Textbook Pages 83 – 84

Have students explain why an object does not belong.

▲ What Shape?

Materials: 3D geometric shapes including cubes, cones, cylinders, and spheres, paper bag or sock

Hide an object in a paper bag or sock. One student feels and describes the solid to his partner, who tries to guess what the shape is.

▲ Domino Count

Materials: Dominoes

Have students count the dots on dominoes and write the corresponding numerals on a whiteboard.

Exercise 1 • page 81

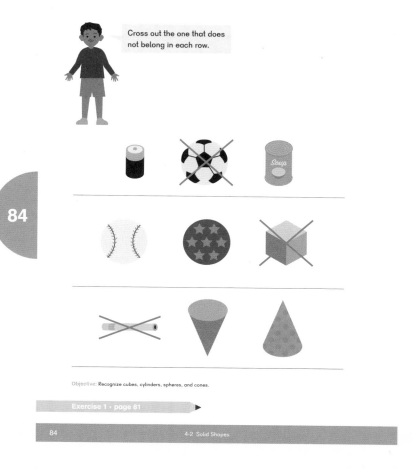

Cross out the one that does not belong in each row.

84

Objective: Recognize cubes, cylinders, spheres, and cones.

Exercise 1 • page 81

84   4-2 Solid Shapes

Extend

★ Make 3D Shapes

Materials: Play dough or salt dough

Have students create 3D geometric shapes using play dough. Students can sort and save their shapes in a 3D classroom "shape museum."

Lesson 3 Closed Shapes

Objective

- Identify closed shapes.

Lesson Materials

- Pipe cleaners or chenille stems, 1 per student

Explore

Pass out pipe cleaners to students and ask them to bend them into shapes.

Ask students to describe their shapes, using words like "straight" and "curved." Ask students if there is an inside and an outside to their shape.

Shapes with an inside and an outside are called closed. Shapes without a definite inside are open.

Have students turn their open shapes into closed shapes.

Learn

Have students identify the shapes on page 85 as open or closed.

Small Group Activity

▲ Draw Shapes

Materials: Art supplies, art paper

Have students draw open and closed shapes, then color in the closed shapes.

Exercise 2 • page 83

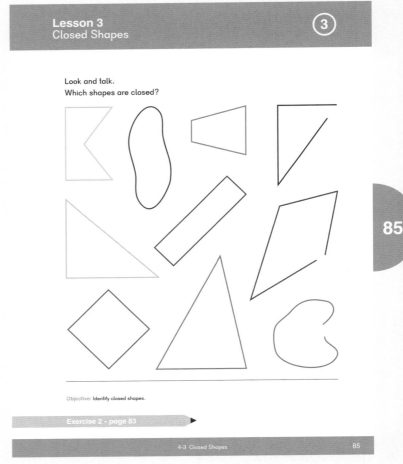

Lesson 3
Closed Shapes
③

Look and talk.
Which shapes are closed?

85

Objective: Identify closed shapes.

Exercise 2 • page 83

4-3 Closed Shapes 85

Extend

★ Geoboards

Materials: Geoboards, rubber bands, Geoboard Dot Paper (BLM)

Have students make closed shapes with rubber bands on their geoboards, then copy the shapes onto the Geoboard Dot Paper (BLM).

Lesson 4 Rectangles

Objective

- Identify rectangles.

Lesson Materials

- Cubes and cuboids from geometric sets or boxes shaped like cubes or cuboids

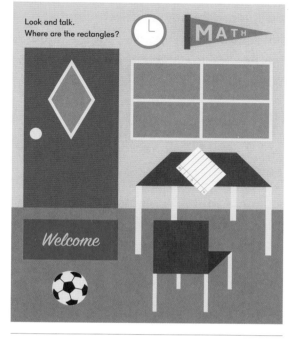

Explore

Provide students with cuboids or boxes and have them describe what the faces look like.

Introduce the term "rectangle" to describe the shape.

Ask students what is the same about each rectangle on the box. Encourage them to notice:

- All rectangles have 4 sides and they all have 4 corners that look the same.
- The sides are straight.
- The sides across from each other are the same.

Send students on a classroom scavenger hunt to find objects that are rectangular or have a rectangular face. Have students share what they've found.

Learn

Have students identify the rectangles on textbook page 86. The table in the textbook is a trapezoid. Some students may see it as a rectangle due to the flat perspective. Have students compare tabletops in the classroom.

Whole Group Activity

▲ Shape Walk

Materials: Sidewalk chalk or large paper rectangles

Draw shapes outside with chalk, or tape down shapes in the classroom. Be sure to include some squares. Have students march around the lines that make the rectangles, making pivots at the corners.

Small Group Activities

▲ **Textbook** Page 87

Students will learn in the next lesson that squares are special rectangles. Discuss that the squares are also rectangles, as they have 4 straight sides and 4 corners that look the same.

▲ **Face Trace**

Materials: 3D objects with rectangular faces, art paper, pencils

Provide students with 3D objects with rectangular faces. Have them trace around a face of a few of the objects and turn them into pictures.

▲ **Roll and Build**

Materials: 6-sided die, linking cubes, Number Cards (BLM) 1 to 10

Provide students with a 6-sided die, linking cubes, and sets of Number Cards (BLM) 1 to 10. Have students order their set of cards. Players take turns rolling the die and choosing a number card on which to build a tower. The roll cannot be split between two numbers. For example, if a player rolls a 3, they can build a tower on the number card for 3 or higher, but not on 1 or 2.

Students must complete a number tower with an exact roll. The first player to build towers for all number cards wins.

Exercise 3 • page 85

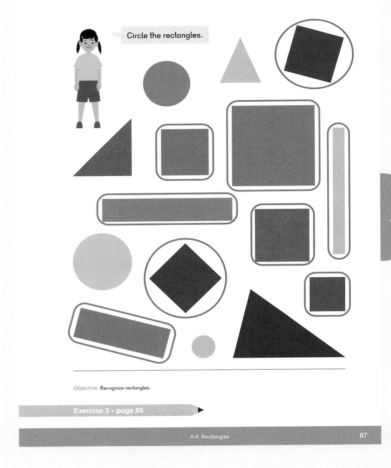

Circle the rectangles.

Objective: Recognize rectangles.

Exercise 3 • page 85

4-4 Rectangles 87

Extend

★ **Not a Rectangle**

Materials: Art paper, pencils

Challenge students to draw four-sided figures that are not rectangles and explain their thinking.

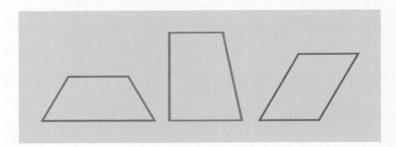

Objective

*   Identify squares.

Lesson Materials

*   Cubes

Explore

Provide students with cubes and have them describe what the faces look like. Ask them what makes these faces a special type of rectangle. Students may describe the shape by saying things like:

*   It has 4 sides.
*   It has 4 corners.
*   All 4 corners look the same.
*   The sides are all straight.
*   All 4 sides are the same.

Tell students that a square is a special rectangle where all sides are the same. Send students on a classroom scavenger hunt to find objects that are square or have a square face. Have students share what they've found.

Learn

Have students identify the squares and contrast them with the other shapes (rectangles, trapezoids, and rhombus) on page 88.

Record students' descriptions of the shapes they see on 3 pieces of chart paper; one labeled Rectangles, one labeled Squares, and one labeled Other 4-sided Shapes.

Lay these charts on the floor so students can use them to sort shapes they make in the **Make and Sort Shapes** on the next page of this Teacher's Guide.

Lesson 5
Squares
⑤

Look and talk.
Where are the squares?

88

Objective: Identify squares.

Exercise 4 · page 87

88   4-5 Squares

Small Group Activities

▲ How Many Can You Make?

Materials: Small square tiles or squares cut from construction paper

Provide students with small squares and have them explore how many different rectangles they can make with 4 small squares, 5, 6, and so on. Students can trace their rectangles onto paper. Alternatively, provide paper squares and have students glue them down.

▲ Make and Sort Shapes

Materials: Sculpting materials such as play dough, Wikki Stix, pipe cleaners, craft sticks, glue, and yarn

Place a variety of sculpting/craft materials in a learning center for students to use to make shapes. Have students sort their shapes on the charts created in **Learn.**

▲ Shape Art

Materials: Sculpting materials, Wikki Stix, pipe cleaners, craft sticks, yarn, glue, large construction or art paper

Students make 4-sided shapes using the materials and glue them to large construction or art paper to make pictures or designs.

As an extension, have students create a graph of the shapes used in their pictures with the labels Rectangles, Squares, and Other.

Include **Roll and Write** to practice number writing.

▲ Roll and Write

Materials: Number Cards (BLM) 1 to 10, 10-sided die or spinner

Provide a 10-sided die, spinner, or Number Cards (BLM) 1 to 10. Students roll the die and practice writing the number five times. Ask them to evaluate each row of numerals and circle the one they feel they wrote best. Roll again and continue.

Exercise 4 • page 87

Extend

★ How Many Squares?

Materials: Geoboards, rubber bands, Geoboard Dot Paper (BLM)

Have students create as many squares as they can on their geoboard. On the Geoboard Dot Paper (BLM), students connect the dots to create as many squares as they can. In either case, squares may overlap.

Teacher's Guide KA Chapter 4

© 2017 Singapore Math Inc.

Lesson 6 Circles and Triangles

Objective

- Identify circles and triangles.

Lesson Materials

- Index cards cut into different triangles

Prior to the lesson, cut index cards into differently shaped triangles. In addition to equilateral and right triangles, provide students with scalene triangles:

Explore

Ask students if they know the name of the shape of the card. If not, introduce the term "triangle," ask students to describe the shape:

- It has 3 sides.
- It has 3 corners.

Send students on a classroom scavenger hunt to find objects that are triangular or have a triangular face. Have students share what they've found.

Learn

Have students identify triangles and circles in the illustration on page 89.

Whole Group Activity

▲ **Shapes of Me (Faces)**

Materials: Paper on which a large circle is drawn, construction paper (rectangles, circles, and triangles, cut out prior to lesson), glue

Provide each student with a piece of paper with a large circle drawn in the middle for a face, and smaller shapes cut out of construction paper for the features: circles, rectangles, squares, and triangles.

Students create a face by gluing shapes onto their circle based on their answers to the following:

- Eyes: Are you a girl or a boy?
  - Girl: circles — Boy: triangles
- Ears: Do you have a pet?
  - Yes: triangles — No: rectangles
- Nose: Do you have any brothers or sisters?
  - Yes: square — No: triangle
- Mouth: How old are you?
  - Glue a circle for each year

Students can finish their face by drawing on hair. Alternatively, students could draw the shapes to create their face.

Small Group Activities

▲ **Textbook** Pages 90 – 91

▲ **Make Shapes**

Materials: Sculpting materials such as play dough, Wikki Stix, pipe cleaners, craft sticks, glue, and yarn

Place a variety of sculpting/craft materials in a learning center for students to use to make shapes. Have students create structures using their shapes. Ask them to describe the shapes used.

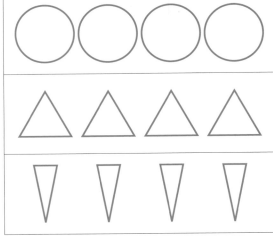

Trace each shape and draw one more.

Objective: Identify, name, and draw circles and triangles.

Color the triangles.

Objective: Recognize triangles as being closed figures with 3 straight sides and 3 corners.

Exercise 5 • page 89

▲ Robots

Materials: Construction paper (rectangles, circles, and triangles, cut out prior to lesson), glue

Use paper cutouts of rectangles (including squares), circles, and triangles to create robot pictures.

▲ Rectangles into Triangles

Materials: Construction paper cut into rectangles

Provide students with paper rectangles and have them cut them in various ways to create triangles.

▲ Shape Game

Materials: One die with 0 instead of 6, one die with shapes: ● – ● – ▲ – ▲ – ▬ – ▬, paper cutouts (circle, triangle, and rectangle-shaped) or pattern block pieces

Students take turns rolling the two dice and collecting the corresponding number of shapes. For example, if they roll a two and a ▬, they collect two rectangles.

The first player to collect 10 of each shape is the winner.

Exercise 5 • page 89 ▶

Extend

★ Pieces of Shapes

Materials: Cardstock or foam, scissors, optional tangram puzzle

Cut squares, rectangles, circles, and triangles out of cardstock or foam. Turn them into tangrams by cutting them into random pieces. Have students put pieces back to form the original shape. Provide a template for the original shape if needed.

Or, provide students with a tangram puzzle. How many different triangles, rectangles, and squares can students make? They can record their discoveries by tracing the tangram pieces.

Lesson 7 Where is It?

Objective

- Use position words to describe the location of an object or shape.

Lesson Materials

- Crayon or other small object for each student

Explore

Provide each student with a classroom object (crayon, marker, etc.) and play a game of **Simon Says**. Use position words (above, below, in front of, behind, etc.) in your commands. For example:

- Simon says, "Put the crayon under your foot."
- Simon says, "Put the crayon above your head."

Learn

Look at textbook page 92. Ask students to explain the location of the objects on the page using position words. Students may say:

- I see a football. The football is under the chair.
- I see a mirror. The mirror is behind the chair and next to the shelf.

Encourage students to use the words "left" and "right" to further describe the position of the objects.

Look and talk.
Use words to describe positions.

92

Objective: Use position words to describe the location of an object.

92   4-7 Where is It?

Small Group Activities

▲ **Textbook** Page 93

▲ **Mirror Me**

Materials: Paper cutouts of circles, triangles, rectangles and squares

Provide students with paper cutouts of circles, triangles, rectangles and squares. Player 1 creates a configuration using four shapes, keeping it hidden from Player 2. Player 1 then describes the configuration using position words, to Player 2, who tries to duplicate it. Player 1 reveals the configuration and Player 2 checks to see if it is a match. Players change roles and play continues.

▲ **Mirror Me — Draw**

Materials: Art paper, markers or crayons

Player 1 draws a simple picture with up to five shapes or items. She describes her picture to Player 2, who tries to draw the same picture with the items in the same relative positions. For example:

- First, draw a black bird in the middle of the page.
- Second, draw an orange circle above the bird.
- Third, draw yellow birdseed to the right of the bird.
- Fourth, draw a blue closed shape to the left of the bird.
- Last, draw green grass below the bird.

▲ **Rainbow Writing**

Materials: Art paper, crayons, markers, or paint

Students practice writing the numerals 2 through 9. Each numeral must be written and traced over with at least five different colors so each numeral resembles a rainbow. For students needing assistance, dotted numerals can be traced.

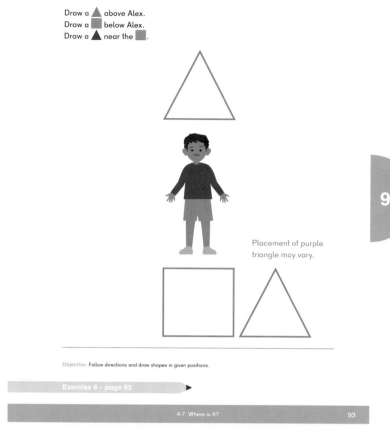

Draw a ▲ above Alex.
Draw a ■ below Alex.
Draw a ▲ near the ■.

93

Placement of purple triangle may vary.

Objective: Follow directions and draw shapes in given positions.

Exercise 6 • page 93

4-7 Where is It? 93

◀ **Exercise 6 • page 93**

Extend

★ **Mirror Me Draw to 10**

Materials: Art paper, art supplies

Player 1 draws a simple picture with up to 10 shapes or items. He describes his picture to Player 2, who tries to draw the same picture.

Lesson 8 Hexagons

Objectives

- Identify and name hexagons.
- Copy a pattern.

Lesson Materials

- Pattern blocks

Explore

Provide students with pattern blocks and have them describe the different shapes.

Note that most pattern block sets consist of six shapes and students are not expected to learn the terms "rhombus" and "trapezoid:"

- Equilateral triangle (Green)
- Hexagon (Yellow)
- Square (Orange)
- Small rhombus (Beige)
- Rhombus (Blue)
- Trapezoid (Red)

Students should describe a hexagon by the number of straight sides. Tell students the name for the 6-sided figure is "hexagon."

Show students an AB pattern with the shapes. Discuss that patterns continue and repeat.

Ask a student to make a different pattern with the blocks. Ensure that the pattern repeats more than once. Have another student describe the pattern. For example:

- Triangle, hexagon, triangle, hexagon ...
- Red, yellow, green, red, yellow, green ...

Repeat this process and have students say the next shape or color in the pattern.

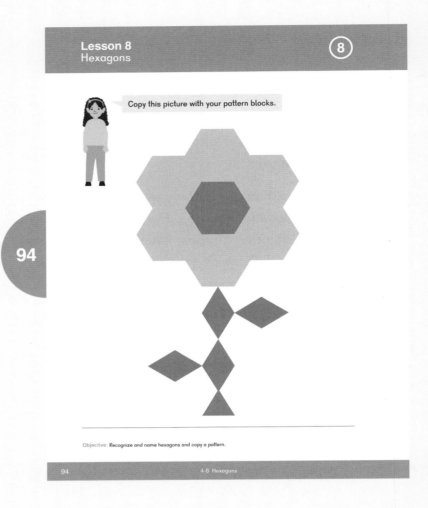

Lesson 8
Hexagons ⑧

Copy this picture with your pattern blocks.

94

Objective: Recognize and name hexagons and copy a pattern.

94   4-8 Hexagons

Learn

Have students make the picture of the flower on page 94 with their pattern blocks and tell how many of each block they used. Encourage the use of position words in their descriptions.

Small Group Activities

▲ **Textbook** Page 95

▲ **What Shape?**

Materials: Pattern block shapes, paper bag or sock

Hide an object in a paper bag or sock. One student feels and describes the shape to his partner, who tries to guess what the shape is.

▲ **Domino Sort**

Materials: Dominoes

Have students take turns choosing a domino and placing it in front of them in order from 0 to 10. If the student already has that number, she returns the domino to the pile. The first player to line up all the numbers 0 through 10 is the winner.

Exercise 7 • page 95

95

Say the name of the shapes, then circle the shape that comes next.

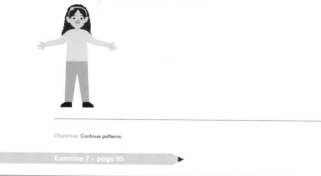

Objective: Continue patterns.

Exercise 7 • page 95

4-8 Hexagons   95

Extend

★ **Symmetry Pictures**

Materials: Pattern blocks

Set up pattern blocks to make the left side (one half) of a picture. Have students make the right side next to the original side.

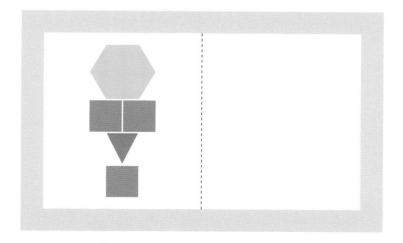

Lesson 9 Sizes and Shapes

Objectives

- Identify shapes and solids according to size.
- Continue a pattern.

Lesson Materials

- Attribute blocks, divided into thick and thin sets
- *Goldilocks and the Three Bears*

Explore

Divide the class into groups of four or five. Provide each group with some attribute blocks that are either thick or thin. Ask the groups to sort their blocks.
Note: Groups may sort by shape or color only and some may sort by size. Ask groups to describe their sort. Then, ask groups to sort their blocks a different way and again describe their sort.

Hold up a large triangle and a small triangle of the same color and thickness. Ask students what makes them the same and what makes them different. When describing the triangles, encourage students to use the phrases "bigger than" and "smaller than."

Learn

Ask groups to sort their blocks by size.

Read *Goldilocks and the Three Bears*. Point out items that are big, bigger, and biggest, and small, smaller, and smallest.

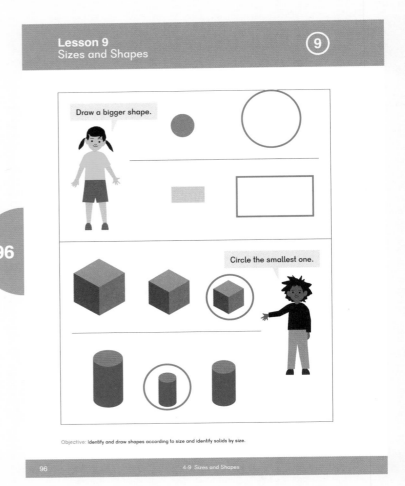

96

Objective: Identify and draw shapes according to size and identify solids by size.

96   4-9 Sizes and Shapes

Teacher's Guide KA Chapter 4   © 2017 Singapore Math Inc.

Whole Group Activity

▲ Shapes & Sizes

Materials: Shapes & Sizes Picture Cards (BLM)

Provide each student with a Shapes & Sizes Picture Card (BLM) showing shapes of various sizes. Ask students to group themselves by shape and order their cards within their groups from smallest to biggest or biggest to smallest. Cards can be collected, shuffled, and passed out again to repeat the activity.

Small Group Activities

▲ Textbook Pages 96 – 97

▲ Roll and Write Patterns

Materials: Number Cards (BLM) 1 to 10, 10-sided die, or spinner

Provide students with a 10-sided die, spinner, or Number Cards (BLM) 1 to 10. Have students roll the die and practice writing the numeral six times, following the pattern big, small, big, small, as they write. Roll again and continue. For example:

▲ Complete the Pattern

Materials: Attribute blocks

Provide pairs of students with attribute blocks. Player 1 makes a pattern with the blocks, including sizes of shapes. Player 2 continues the pattern by adding the next shape in the pattern. Players switch roles and play continues.

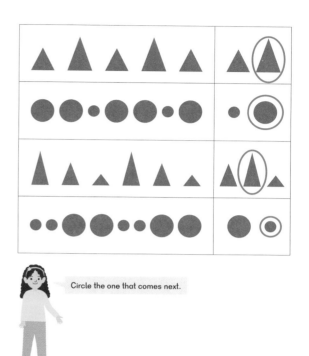

Circle the one that comes next.

Objective: Continue patterns.

Exercise 8 • page 97

4-9 Sizes and Shapes 97

Extend

★ Attribute Trains

Materials: Attribute blocks

Pass out 5 attribute blocks randomly to two or more players. Player 1 places a block down as the first car of the train. Player 2 places a block down as the second car; his block must have at least one same attribute as the previous block. If a player cannot make a play, he must pick a block from the draw pile. The first player to play all of his blocks wins.

For more of a challenge, have students play a block that has only one attribute the same as the previous block.

Objective

- Make new shapes by combining basic shapes.

Lesson Materials

- Pattern blocks

Explore

Provide pairs of students with triangle, square, and hexagon pattern blocks. Ask the pairs of students to create as many different shapes as they can using only 5 pattern blocks. Students can trace their shapes on paper. Give them a few minutes to compete their task, then ask them to describe to their partner the blocks they used. Encourage students to use color, the shape names, and position words in their descriptions.

Learn

Look at page 98 and discuss Alex's and Emma's shapes. Using the same blocks Alex used, ask students to make a different shape than Alex. Ask the same of Emma's shape. Have students share their shapes.

Lesson 10
Combine Shapes ⑩

Look and talk.
What shapes are Alex and Emma using to make other shapes?

98

Objective: Make new shapes by combining basic shapes.

98   4-10 Combine Shapes

Small Group Activities

▲ **Textbook** Page 99

▲ **Mirror Me — Pattern Blocks**

Materials: Paper cutouts of circles, triangles, rectangles, and squares

Player 1 creates a shape using 5 pattern blocks, keeping it hidden from Player 2. Player 1 then describes the shape to Player 2 using shape names and position words. Player 2 attempts to duplicate the shape. Player 1 reveals the shape and if Player 2 was able to duplicate it, Player 2 scores a point. Players switch roles and play continues.

▲ **Pattern Block Pictures and Shapes**

Materials: Tangram Pieces (BLM)

Have students make different pictures using Tangram Pieces (BLM). Allow students to use all the tangram pieces, even the beige and blue ones.

★ For more challenge, ask students to make specific shapes such as a large hexagon.

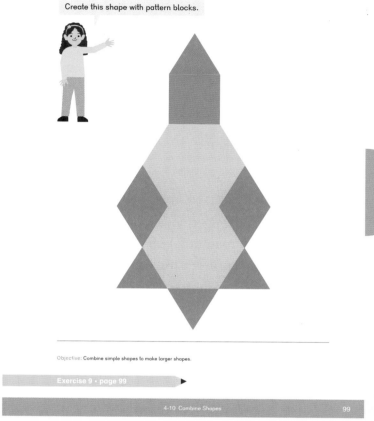

Create this shape with pattern blocks.

Objective: Combine simple shapes to make larger shapes.

Exercise 9 • page 99

4-10 Combine Shapes   99

Exercise 9 • page 99 ▶

Extend

★ **Pattern Blocks**

Materials: Pattern blocks

Provide students with all the shapes in a pattern block set and challenge them to make the larger shapes, like the hexagon, out of the smaller squares and triangles. Have them discuss the relationships between the smaller shapes and the larger shape. A student might respond, "I used 6 triangles to make a hexagon."

Lesson 11 Graphs

Objective

- Represent data in a graph.

Lesson Materials

- 3D shapes from home
- Bar graph large enough to hold shoes, made with poster paper or tape on a rug, with columns labeled with the names of the 3D shapes learned in Lessons 1 and 2

Explore

Ask students to choose a 3D object from objects brought from home or from the "shape museum."

Sort the objects by shape. Once sorted, ask students what could be done to organize the shapes to make it easier to compare the quantity in each group.

Organize the shapes onto a large bar graph. Ask students to suggest a title for the graph. Ask:

- Which shape do we have the most of? (A column with no shapes is actually the least. Review the concept of 0.)
- Are there the same amount of any?

Learn

Look and talk about the graph on page 100. Ask:

- What is the title of the graph?
- How will we make sure we've included all the shapes in the graph?

Share strategies for making sure all the shapes are included in the graph. Students might suggest that the shapes are crossed off as they are counted and colored in on the graph.

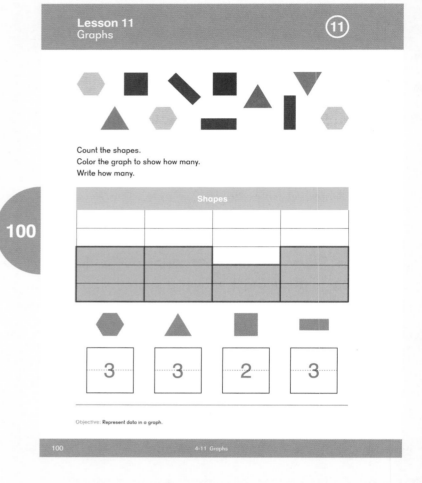

Lesson 11
Graphs
(11)

Count the shapes.
Color the graph to show how many.
Write how many.

Shapes			

3	3	2	3

Objective: Represent data in a graph.

100   4-11 Graphs

Whole Group Activity

▲ Class Graphing

Materials: Blank Graph (BLM)

Build a classroom graph to compare answers to a question that has meaning to your class. For example:

- What is your favorite snack?
- What is your favorite recess game?
- Which greeting or activity should we do at morning meeting tomorrow?
- Which book should we read tomorrow?

Discuss labels and a title that would be appropriate for your graph.

Small Group Activities

▲ **Textbook** Pages 100 – 101

▲ **Build and Graph**

Materials: Blank Graph (BLM), pattern blocks

Have students create a shape no larger than an $8\frac{1}{2} \times 11$ sheet of paper with pattern blocks of various shapes.

Using a Blank Graph (BLM), label the graph with pictures of a triangle, square, hexagon, and rectangle along the bottom row. Have students count the shapes used and complete the graph.

▲ **Roll and Graph**

Materials: 6-sided die, 6 × 10 grid with the numbers 1 through 6 written beneath each respective column

Provide students with a 6-sided die and a 6 × 10 grid as described in **Materials**. Students roll the die and write the number above the column label to build a graph of numbers. For example, a student rolls a 5, finds the 5 column and writes the numeral 5 in the lowest available space. Students continue to roll and write the numbers to build the graph.

Exercise 10 • page 101 ▶

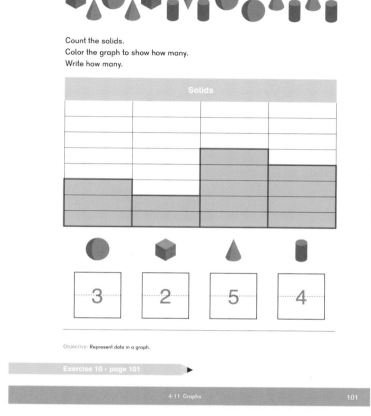

Count the solids.
Color the graph to show how many.
Write how many.

		Solids		

3 2 5 4

Objective: Represent data in a graph.

Exercise 10 • page 101 ▶

4-11 Graphs   101

Extend

★ **How Many Shapes?**

How many squares are in this figure?

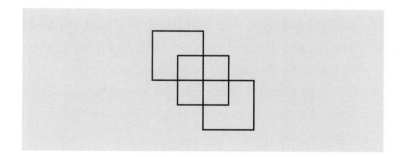

How many triangles are in this figure?

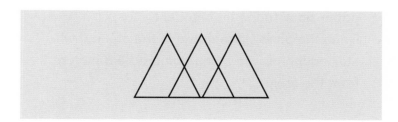

Teacher's Guide KA Chapter 4

Lesson 12 Practice

Objective

- Practice concepts from the chapter.

After students complete the **Practice** in the textbook and workbook, have them continue identifying two-dimensional and 3D shapes by creating and extending patterns using **Activities** from the chapter.

Whole Group Activity

▲ **Pass the Paper**

Materials: Blank paper for each student, pencils

Each student begins with a piece of paper. Ask students to draw a shape of their choosing, then have them pass the paper to another student. Each student can continue adding shapes and passing to a new person. If desired, the teacher can suggest different shapes or additions to the drawing with each passing of the paper.

After the paper has been drawn on by four or five students, have all papers returned to the original students. Have each student try to make a picture from the shapes on her paper.

Example:

- The first student puts her name on the paper and draws a triangle, then passes to the next student.
- The second student draws four dots on the paper and passes the paper to another student.
- The third student is asked to connect the previous student's dots to make a shape, then pass the paper again.
- Next, ask the fourth student to draw a rectangle (or square, or 4-sided shape that isn't a rectangle, an open shape, etc.) and pass the paper back to the first student.
- Once the paper is returned to the student who started the drawing, she can create a picture from the shapes.

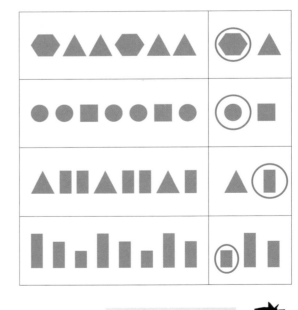

136   Teacher's Guide KA Chapter 4   © 2017 Singapore Math Inc.

Extend

★ Tangrams

Materials: Tangram Pieces (BLM), copied onto cardstock, *Three Pigs, One Wolf, and Seven Magic Shapes* by Grace Maccarone

Tangrams can be used to develop problem solving and logical thinking skills, visual-spatial awareness, creativity, and many mathematical concepts.

The book *Three Pigs, One Wolf, and Seven Magic Shapes* is an introduction to tangrams for younger students.

Consider providing outlines for students to match the shapes when making a tangram picture. You can print the Tangram Pieces (BLM) on cardstock and cut them out if needed.

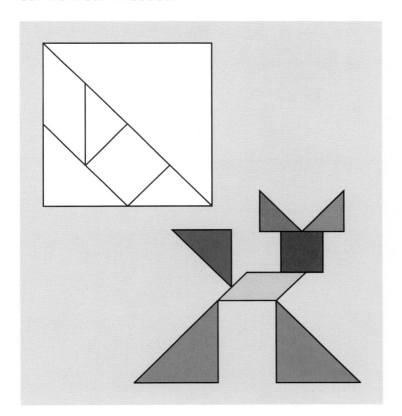

104

Count the solids.
Color the graph to show how many.
Write how many.

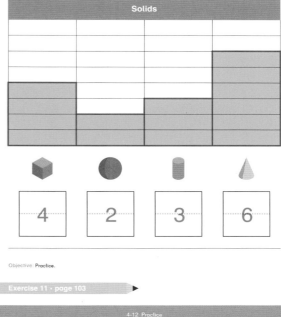

Solids			

4   2   3   6

Objective: Practice.

Exercise 11 • page 103

104   4-12 Practice

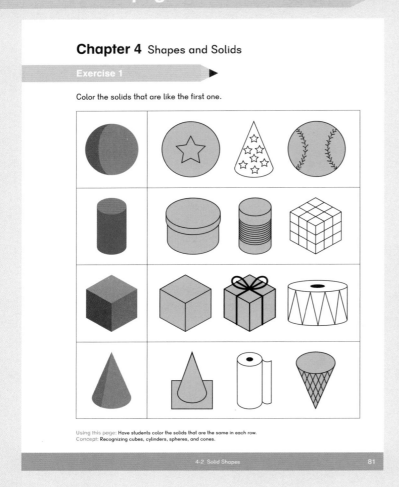

Chapter 4 Shapes and Solids

Exercise 1

Color the solids that are like the first one.

Using this page: Have students color the solids that are the same in each row.
Concept: Recognizing cubes, cylinders, spheres, and cones.

4-2 Solid Shapes   81

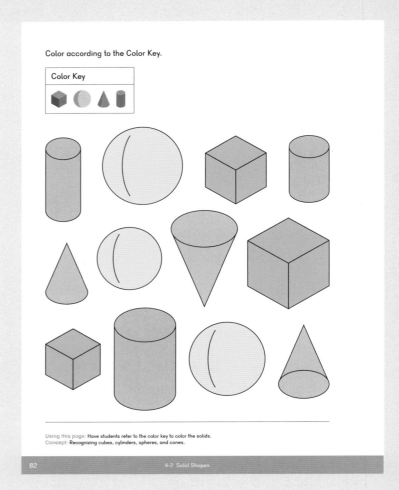

Color according to the Color Key.

Color Key

Using this page: Have students refer to the color key to color the solids.
Concept: Recognizing cubes, cylinders, spheres, and cones.

82   4-2 Solid Shapes

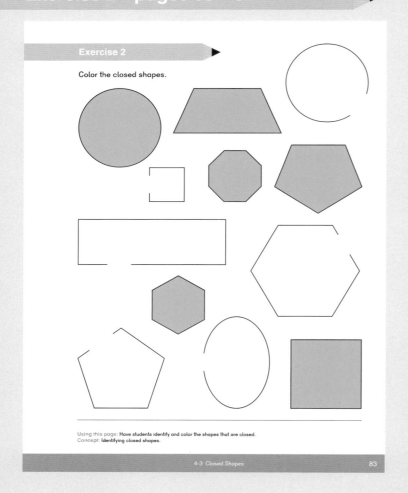

Exercise 2

Color the closed shapes.

Using this page: Have students identify and color the shapes that are closed.
Concept: Identifying closed shapes.

4-3 Closed Shapes   83

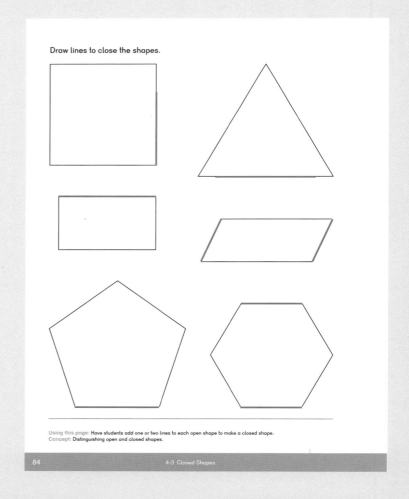

Draw lines to close the shapes.

Using this page: Have students add one or two lines to each open shape to make a closed shape.
Concept: Distinguishing open and closed shapes.

84   4-3 Closed Shapes

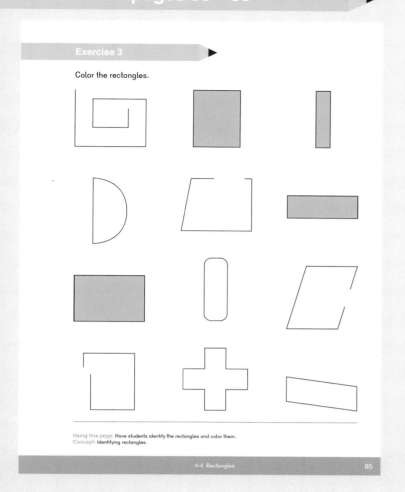

Exercise 3

Color the rectangles.

Using this page: Have students identify the rectangles and color them.
Concept: Identifying rectangles.

4-4 Rectangles   85

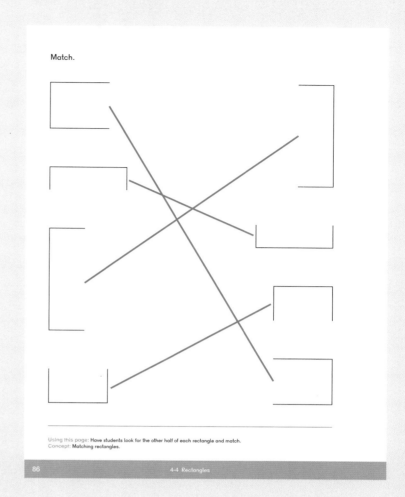

Match.

Using this page: Have students look for the other half of each rectangle and match.
Concept: Matching rectangles.

86   4-4 Rectangles

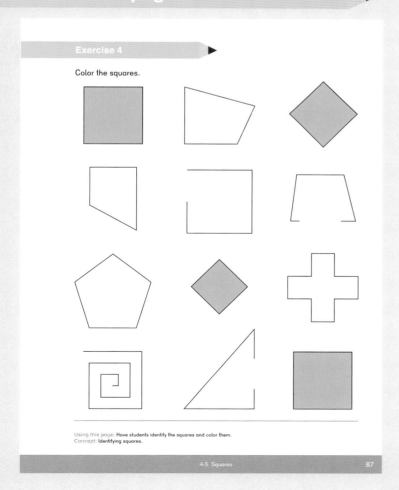

Exercise 4

Color the squares.

Using this page: Have students identify the squares and color them.
Concept: Identifying squares.

4-5 Squares   87

Color the squares.

Using this page: Have students identify the squares and color them.
Concept: Identifying squares.

88   4-5 Squares

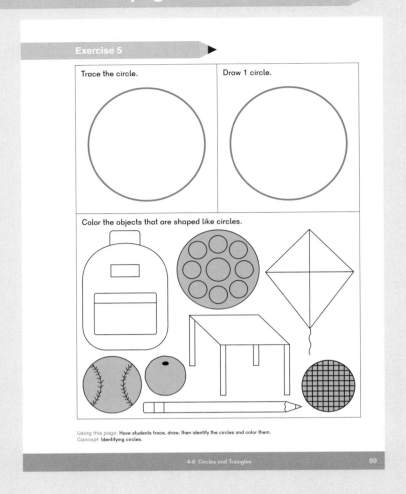

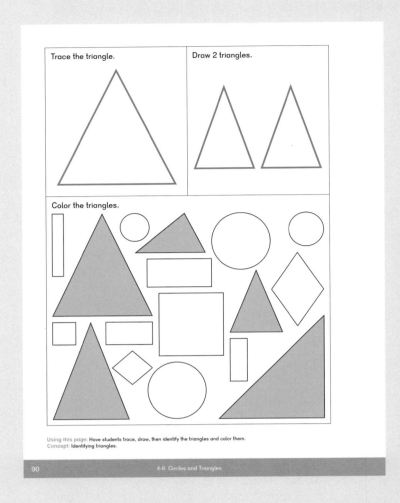

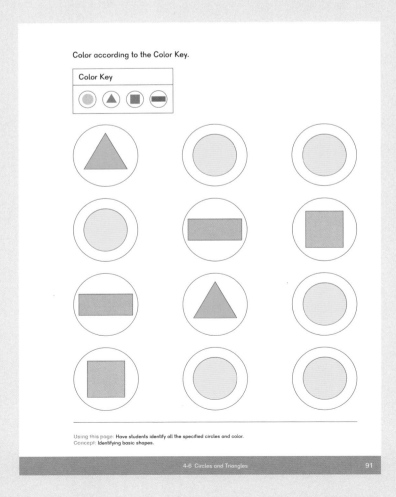

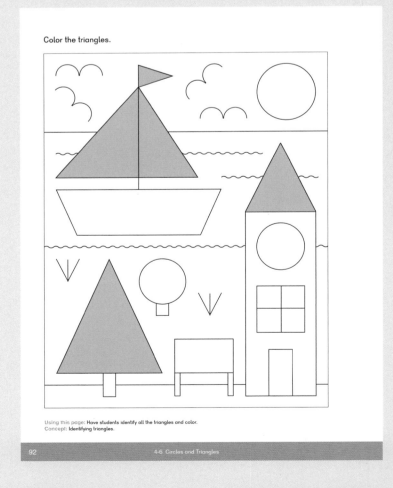

Exercise 6

Follow the directions and color.
Color the ⬡ at the top 🔴.
Color the ⬡ below the red cube 🔴.
Color the ◯ next to the green cylinder 🟢.
Color the ⬡ between the cube and sphere 🔵.

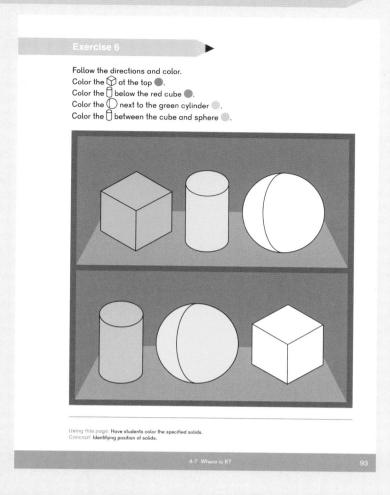

Using this page: Have students color the specified solids.
Concept: Identifying position of solids.

4-7 Where is It?   93

Follow the directions and paste the shapes.
Paste the ▲ above the ●.
Paste the ▬ below the ●.
Paste each ◯ on the ▬.
Paste the ▬ on the left of the ▬.
Paste the ▬ on the right of the ▬.
Paste the ■ below the ▬.

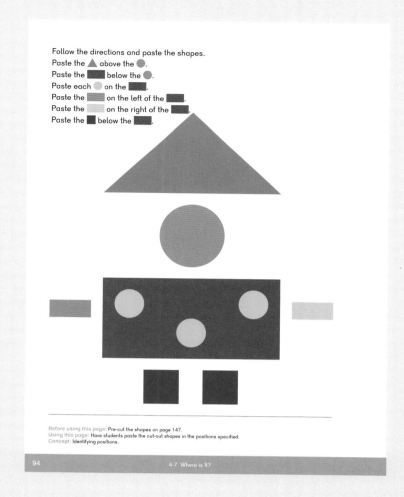

Before using this page: Pre-cut the shapes on page 147.
Using this page: Have students paste the cut-out shapes in the positions specified.
Concept: Identifying positions.

94   4-7 Where is It?

Exercise 7

Color the hexagons to form a path from the bee to the hive.

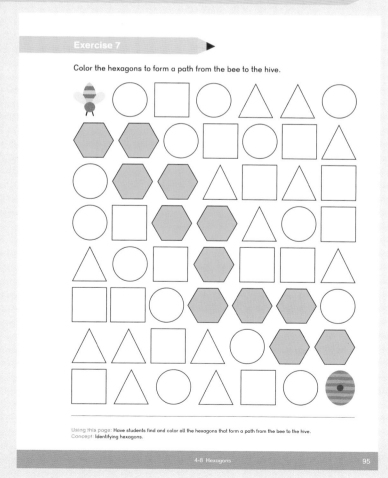

Using this page: Have students find and color all the hexagons that form a path from the bee to the hive.
Concept: Identifying hexagons.

4-8 Hexagons   95

Draw and color the correct shapes missing in each pattern.

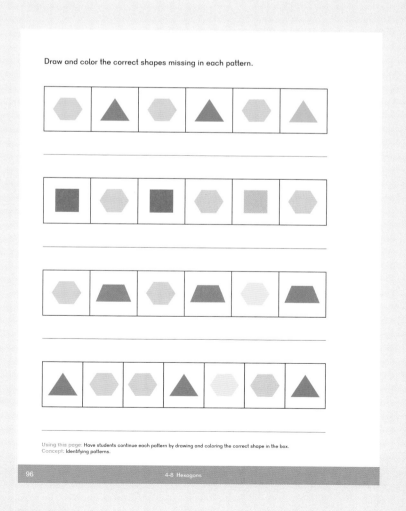

Using this page: Have students continue each pattern by drawing and coloring the correct shape in the box.
Concept: Identifying patterns.

96   4-8 Hexagons

Teacher's Guide KA Chapter 4   141

Exercise 8

Color the shapes of the same size.

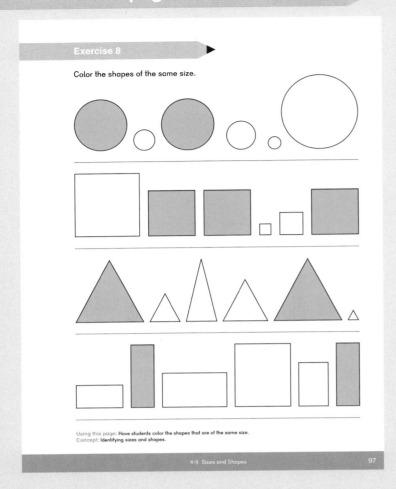

Using this page: Have students color the shapes that are of the same size.
Concept: Identifying sizes and shapes.

4-9 Sizes and Shapes   97

Draw the shape that completes the pattern.

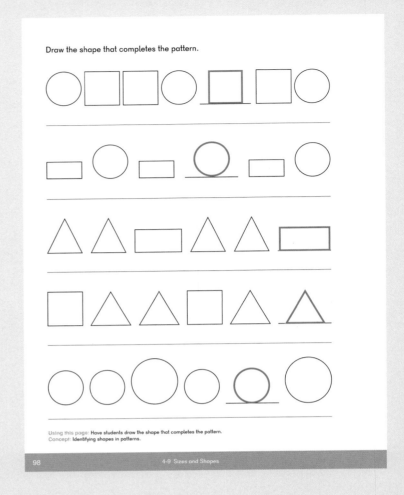

Using this page: Have students draw the shape that completes the pattern.
Concept: Identifying shapes in patterns.

98   4-9 Sizes and Shapes

Exercise 9

Copy the picture and write the number.

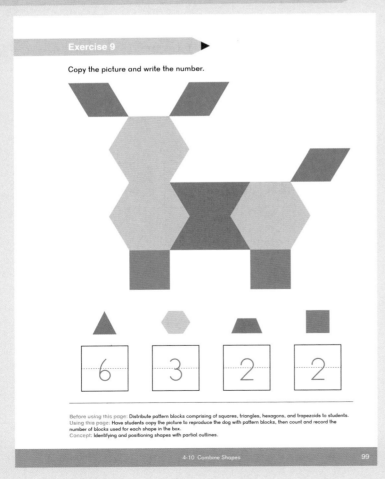

Before using this page: Distribute pattern blocks comprising of squares, triangles, hexagons, and trapezoids to students.
Using this page: Have students copy the picture to reproduce the dog with pattern blocks, then count and record the number of blocks used for each shape in the box.
Concept: Identifying and positioning shapes with partial outlines.

4-10 Combine Shapes   99

Create your own shapes picture!

Before using this page: Pre-cut the shapes on page 149.
Using this page: Have students create their own picture using the shape cut-outs.
Concept: Creating with shapes.

100   4-10 Combine Shapes

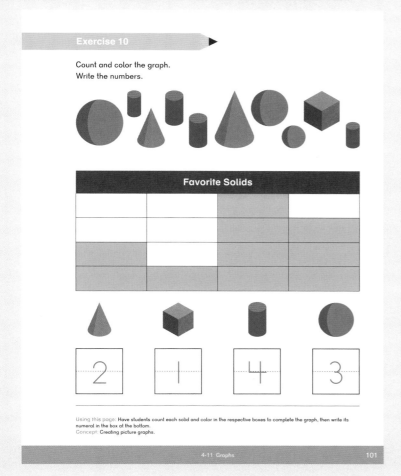

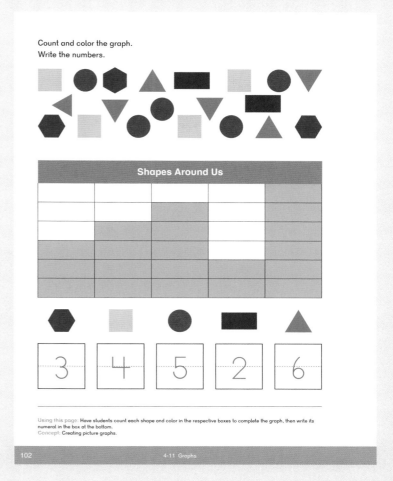

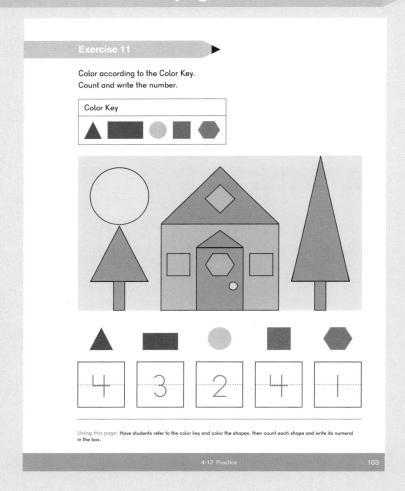

Exercise 11

Color according to the Color Key.
Count and write the number.

Color Key

Using this page: Have students refer to the color key and color the shapes, then count each shape and write its numeral in the box.

4-12 Practice   103

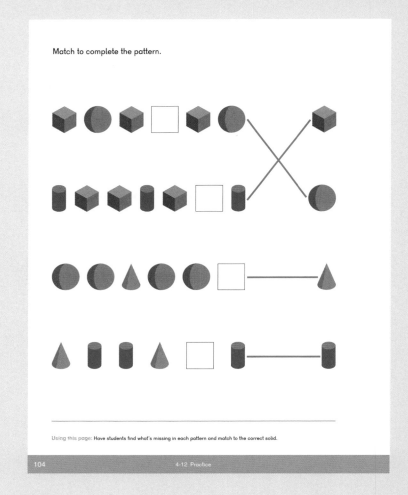

Match to complete the pattern.

Using this page: Have students find what's missing in each pattern and match to the correct solid.

104   4-12 Practice

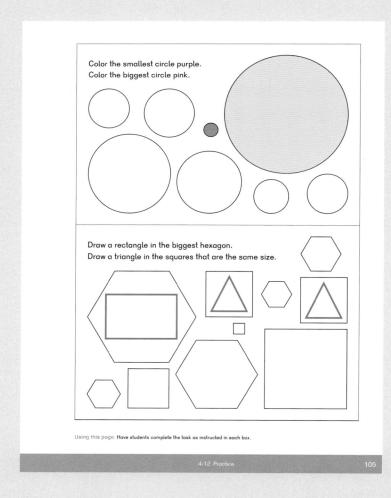

Color the smallest circle purple.
Color the biggest circle pink.

Draw a rectangle in the biggest hexagon.
Draw a triangle in the squares that are the same size.

Using this page: Have students complete the task as instructed in each box.

4-12 Practice   105

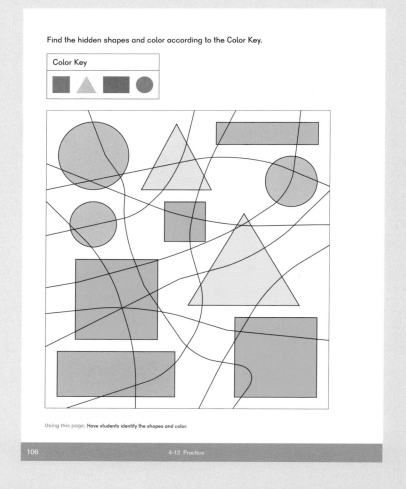

Find the hidden shapes and color according to the Color Key.

Color Key

Using this page: Have students identify the shapes and color.

106   4-12 Practice

Suggested number of class periods: 10 – 11

Lesson		Page	Resources		Objectives
	Chapter Opener	p. 149	TB:	p. 105	
1	Comparing Height	p. 150	TB: WB:	p. 106 p. 107	Compare height.
2	Comparing Length	p. 153	TB: WB:	p. 109 p. 109	Compare length.
3	Height and Length — Part 1	p. 156	TB: WB:	p. 112 p. 111	Compare height and length indirectly.
4	Height and Length — Part 2	p. 158	TB: WB:	p. 114 p. 113	Measure objects using non-standard units.
5	Weight — Part 1	p. 160	TB: WB:	p. 116 p. 117	Compare weight.
6	Weight — Part 2	p. 162	TB: WB:	p. 118 p. 119	Compare weight.
7	Weight — Part 3	p. 163	TB: WB:	p. 119 p. 121	Measure weight with non-standard units.
8	Capacity — Part 1	p. 165	TB: WB:	p. 121 p. 123	Compare capacity.
9	Capacity — Part 2	p. 167	TB: WB:	p. 123 p. 125	Measure capacity with non-standard units.
10	Practice	p. 169	TB: WB:	p. 125 p. 127	Practice concepts from the chapter.
	Workbook Solutions	p. 171			

Measurement adds real-life context to math content and gives children opportunities to estimate and visualize, both key skills in mathematical understanding. Measurement also provides students the chance to practice counting and computational skills.

In **Dimensions Math® Pre-Kindergarten A**, students investigated height, length, and weight. In this chapter, students will compare attributes of items using direct and indirect measurement. They will define measurement as the size, amount, or degree of something by using an instrument or device marked in standard units or by comparing it with an object of known size.

Direct measurement refers to measuring exactly the thing that you're looking to measure.

Indirect measurement means that you're measuring one thing by measuring something else. For example:

- If a ribbon the length of a table is held up to the length of the door and is longer than the door, we can say that the table is longer than the door.
- Water is poured from one full pitcher into a bowl and overflows the bowl. Water is poured from a tall thin beaker into the same bowl and does not fill the bowl. We can say that the pitcher holds more water than the beaker.

This chapter is meant to help students recognize the attributes of length (including height), weight, and capacity. Students will begin to compare these attributes using ribbons, body parts, and linking cubes.

Many students may mistakenly assume that a larger object is also a heavier object. Ensure students have a chance to physically explore this feature of weight by specifically choosing objects of different densities to show that sometimes smaller objects can weigh more.

Capacity is the amount of liquid a vessel can hold, while volume is the amount of space the liquid takes up. Students will compare capacity of cups of different shapes. Encourage students to realize that taller cups do not necessarily have greater capacity.

While some students may already know how to properly compare using words like "tall," "short," and, "long," this chapter is meant to solidify those ideas. Emphasis should be placed on understanding the joint relationships expressed. That is, if a bottle is taller than a cup, the cup is shorter than the bottle.

When objects are measured using units, they often do not measure to an exact whole unit. Students will have to approximate which whole unit the object is closest to. Students should be encouraged to express the unit of measurement whenever measuring. When using linking cubes to measure the height of a bottle, for example, students should answer, "The bottle is ____ linking cubes high."

Storybooks that go along with content in this chapter can be added as time allows or for additional exploration. Storybook suggestions are listed on the following page.

Materials

- Linking cubes
- Ribbon or string
- Straws
- Toothpicks and mini-marshmallows (or clay)
- Balance scales
- Balloons
- Rocks of varying weights, including pumice
- **Capacity items used to measure:**
  - Water
  - Sand
  - Rice
  - Beans
  - Seeds (birdseed or flax seed works well)
  - Water beads
- **Containers:**
  - Water bottles
  - Plastic cups
  - Plastic food containers
  - Pans and pots
  - Muffin tins
  - Jugs
  - Drink bottles
  - Snack containers
  - Small drinking cups
  - Plastic beakers or cylinders
  - Plastic tubs
  - Pails
  - Ice cube trays

Blackline Masters

- Comparing Height
- Straw Sort
- Comparing Length
- Comparing Weight
- Die

Storybooks

- *Tall* by Jez Alborough
- *Who Sank the Boat?* by Pamela Allen
- *Actual Size* by Steve Jenkins
- *Guess How Much I Love You* by Sam McBratney
- *How Deep is the Sea?* by Anna Milbourne
- *How High is the Sky?* by Anna Milbourne
- *Mighty Maddie* by Stuart J. Murphy
- *How Big is a Foot?* by Rolf Myller
- *Yertle the Turtle and Other Stories* by Dr. Seuss
- *Drat That Fat Cat!* by Pat Thomson
- *Balancing Act* by Ellen Stoll Walsh
- *Goldilocks and the Three Bears*

Letters Home

- Chapter 5 Letter

Notes

Chapter Opener

Lesson Materials

- Heavy and light rocks, including pumice

Chapter 5

Compare Height, Length, Weight, and Capacity

105

105

This **Chapter Opener** is designed to help students realize that in describing the world around them, we often refer to the length, weight, or capacity of objects, and compare those attributes. One child is taller than another, one cup holds more juice than another, etc. Students can then begin to think about ways to compare these attributes more precisely.

Generally, the word "height" is used for the vertical length of an object, or its distance from the ground (e.g. the height of an airplane), and "length" for horizontal length. Students will explore the length of objects that can be reoriented to have height. Height, width, depth, etc., are all lengths.

Engage students in a discussion of what they know about height, length, and weight. Ask students to find a tall item and a short item in the classroom, a long item and a short item, and a heavy item and a light item. Look and talk about the objects on page 105.

This discussion is informal. Students are not expected to recognize the interchangeability of these words for some situations. Students should begin to recognize that a large item is not necessarily heavier than a smaller item.

Extend

★ Rock Talk

Materials: Rocks of varying weight, pumice

Have students share what they know about rocks and then explore and describe examples of the provided rocks.

Lesson 1 Comparing Height

Objective

- Compare height.

Explore

Have students partner up and compare their heights. Who is taller? Who is shorter? Divide the class into groups of 4 or 5 and have students compare their heights again. Challenge students to order themselves from shortest to tallest without speaking.

Working with one group, have one of the shorter students stand on a step stool to now be above the tallest student in the group.

Ask if this student's height has changed. Tell students that we can't compare heights unless the objects being compared are on the same starting point.

Learn

Show two objects of different heights that are not lined up at the same baseline or starting point. Ask if it is easier or more difficult to compare objects that are not lined up. Discuss the importance of comparing height from the same starting point.

Discuss pages 106 and 107.

Whole Group Activity

▲ Tall & Short Hunt

Have students look around the classroom to find two objects to bring back to the group. Ask students to compare the heights of their objects, using the phrases "taller than" and "shorter than."

Ask students to partner up and organize their four objects from shortest to tallest. Switch partners and repeat.

Lesson 1
Comparing Height ①

Look and talk.
Who is taller?

Mei   Emma

Objective: Compare height.

106   5-1 Comparing Height

106

Small Group Activities

▲ **Textbook** Page 108

▲ **Which is Taller?**

Materials: Comparing Height (BLM)

Allow students to explore the classroom, comparing the heights of objects. For example, students could compare the height of the door to the height of a desk. Students can draw the objects being compared on Comparing Height (BLM).

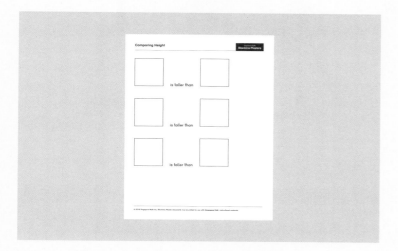

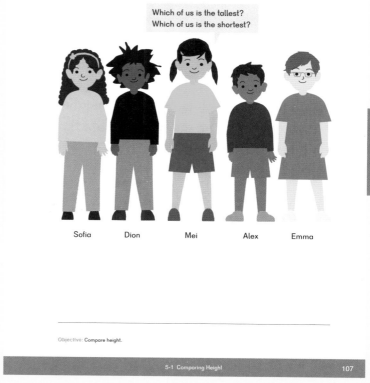

Which of us is the tallest?
Which of us is the shortest?

Sofia   Dion   Mei   Alex   Emma

Objective: Compare height.

5-1 Comparing Height 107

107

▲ **How High Can You Build It?**

Materials: Toothpicks or craft sticks, clay, mini marshmallows or clothespins

Using toothpicks or craft sticks, and clay, mini marshmallows, or clothespins as joiners, challenge students to build the tallest tower.

▲ **Build It**

Materials: Number Cards (BLM) 1 to 10, blocks or linking cubes, two-color counters

Students work in groups of two to four.

Each player draws a Number Card (BLM) 1 to 10 and builds a tower with the number of blocks or linking cubes to match the number shown on the card. Students then compare the heights of their towers and flip a two-color counter. Red means "find the tallest tower," and white (or other side color) means "find the shortest tower."

The player with the corresponding tower (either tallest or shortest, depending on flip) scores a point.

Extend

★ **Line Up**

Materials: Toys, building blocks or boxes

Have students create a line of varying height dolls or toys that all appear to be the same height by placing them on building blocks or boxes.

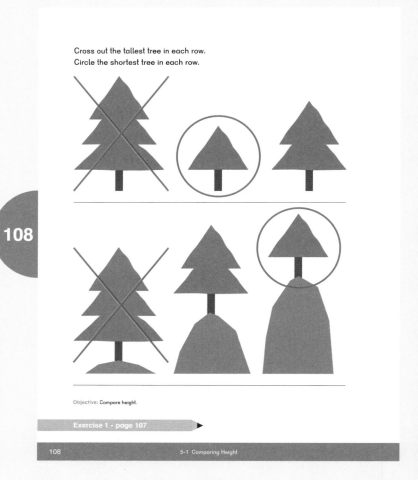

Cross out the tallest tree in each row.
Circle the shortest tree in each row.

Objective: Compare height.

Exercise 1 • page 107

108   5-1 Comparing Height

Objective

• Compare length.

Look and talk.
Which is longest?
Which is shortest?

109

Objective: Compare length.

5-2 Comparing Length   109

Explore

Ask students to find two objects in the classroom of different lengths. Have students share their findings using the phrases "longer than" and "shorter than." Add a third item and ask students, "Which is the longest of the three items? Which is the shortest?"

Learn

Discuss the height and length of various objects. Ask students if the rug has height (it may have a small amount of height) or length, the height or length of a door, etc. Discuss how some objects can have height and length.

Show two objects of different lengths that are not lined up at the same baseline or starting point. Discuss how it is easier to compare length when objects have the same starting point.

Look at page 109 and discuss the illustration. Students should notice that for the lizards, the starting point is on the left, but the sticks are lined up to the right.

Point out to students that we might say the sun is high in the sky, or the mountain on the right is the highest or tallest, but we use the term "tall" (not "high") for people.

Whole Group Activity

▲ Long & Short Hunt

Have students look around the classroom for two objects that have length, and bring them back to the group. Ask students to compare the lengths of their objects, using the phrases "longer than" and "shorter than." Ask students to partner up and organize their four objects from shortest to longest. Switch partners and repeat.

Small Group Activities

▲ Textbook Pages 110 and 111

▲ Straws

Materials: Straws, Straw Sort (BLM)

Activity 1: Cut a set of straws to match Straw Sort (BLM). Have students match straws by length to the lines on the worksheet.

Activity 2: Mix up the straws. Have students put straws in order from shortest to longest or longest to shortest. Pool noodles would also work for this activity.

Activity 3: Have students find objects in the classroom that are the same length as the set of straws.

Activity 4: One student selects a quantity of straws and puts them in his/her hand, showing the ends of the straws to be the same length. Players take turns drawing straws. The player who draws the shortest straw wins.

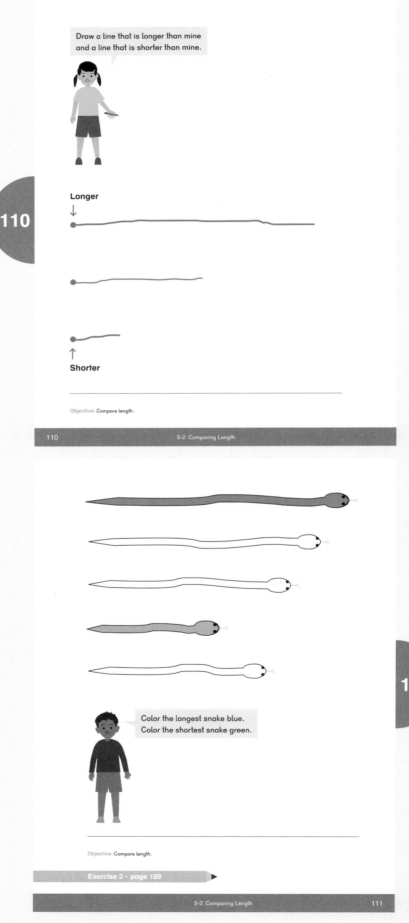

▲ Feather in Your Cap

Materials: Yankee Doodle (VR), feathers, glue, art paper, crayons or markers

Teach students the song "Yankee Doodle" (VR). Have stations set up with feathers of different lengths. Have students order feathers by length. Give each student a piece of paper and a glue stick. Have students draw a hat, then use the feathers to decorate it.

▲ Play Dough Snakes

Materials: Clay or play dough

Provide clay or dough for students to use to create snakes. Have them compare the lengths of their creations. To add a challenge, create and show them a few snakes that are not stretched into a straight line. Ask students to tell which snakes created by you are longer and shorter than their snakes.

Exercise 2 • page 109

Extend

★ How Long is the Hall?

Have students work with partners. Students will use their arm spans or lengths of their bodies to measure the length of the hall or classroom. Partner groups can compare their findings. Why might they be different?

★ Measuring Feet

Materials: Ribbons or yarn

Provide students with ribbon or yarn. Students work with a partner to measure the length of each other's feet. Display the lengths of ribbons on a wall or bulletin board. These can be used again for measuring with non-standard units in a future lesson.

Lesson 3 Height and Length — Part 1

Objective

- Compare height and length indirectly.

Lesson Materials

- Cardboard strips or lengths of ribbon to match the length of immovable objects in the classroom as pictured in the textbook on page 112
- *How Big is a Foot?* by Rolf Myller

Explore

Pass out a cardboard strip or ribbon to pairs of students. Have each pair find a classroom object that is the same length or height as their piece of cardboard or ribbon. Have students share the object. Have two groups compare the lengths of their objects by comparing their lengths of cardboard or ribbon.

For example, the bookshelf is longer than the desk, because the ribbon that is the same length as the bookshelf is longer than the one that is the same length as the desk.

Learn

Look at page 112 and discuss the illustration. Ask students to call out the objects being measured from shortest to longest. Ask students how they know. Record the order on the board or chart paper.

Whole Group Activities

▲ **Textbook** Before looking at page 13, read the book *How Big is a Foot?* by Rolf Myller aloud to students.

Then, have them look at page 113. Ask them what would happen if Dion measured heel to toe and recorded a measurement for the sidewalk crack, and then Emma, Sofia, Alex, and Mei did the same. Do students think that all of the measurements would be the same? Have them explain their reasoning.

Look and talk.

Objective: Compare length of objects indirectly using a third object.

112   5-3 Height and Length — Part 1

We can measure with our feet.

Objective: Measure objects using non-standard units.

Exercise 3 · page 111

5-3 Height and Length — Part 1   113

▲ How Big is Your Foot?

Materials: Art paper, scissors

Have students trace their feet on paper, cut the shapes out, and use them to measure objects in the room. Be sure that students refer to the length of the object with the unit "_____'s feet." For example, "The rug is 8 Sam's feet long."

Small Group Activities

▲ Which is Longer?

Materials: Comparing Length (BLM), cardboard strips or ribbon

Allow students to explore the classroom, comparing the lengths of objects by using their cardboard strip or ribbon. For example, students could compare the length of a book to the length of a pencil. Students can draw objects being compared on Comparing Length (BLM).

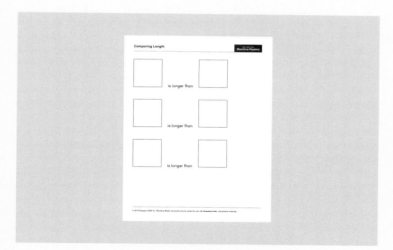

▲ Make a Shape That is ...

Materials: Geoboards, rubber bands

Have students use geoboards and rubber bands to make shapes that are taller or longer and shorter. For example, make a triangle that is taller than a square. Review triangles, squares, and rectangles.

Take it Outside

▲ Long Jump

Materials: Long jump pit (if possible), measuring tape or ribbon, chalk

Students can measure their jumps with ribbon in a long jump pit or from a chalk line for a standing long jump.

Exercise 3 • page 111

Extend

★ If I Were as Tall as a ...

Materials: Recording device, art paper, markers or crayons

Have students write (or record) a story, "If I were as tall (or short) as a _____." Allow them to fill in the blank. Encourage them to illustrate their story.

Objective

- Measure objects using non-standard units.

Lesson Materials

- Ribbons or pieces of yarn, cut to various lengths, 2 per pair of students

Explore

In partner groups, have students compare the length of two pieces of ribbon. Have them measure the length of each piece of ribbon using linking cubes. Ask, "What do you notice about the length of the ribbon and the number of cubes needed?"

For example, a student may answer, "The longer ribbon needed more cubes than the shorter ribbon."

Learn

Look at page 114 and discuss the illustration. Ask students to identify the longest ribbon and the shortest ribbon. Ask, "Do you think the red ribbon or the blue ribbon will need more cubes? How many more?" Compare the yellow ribbon to the green ribbon.

Small Group Activities

▲ **Textbook** Pages 114 – 115

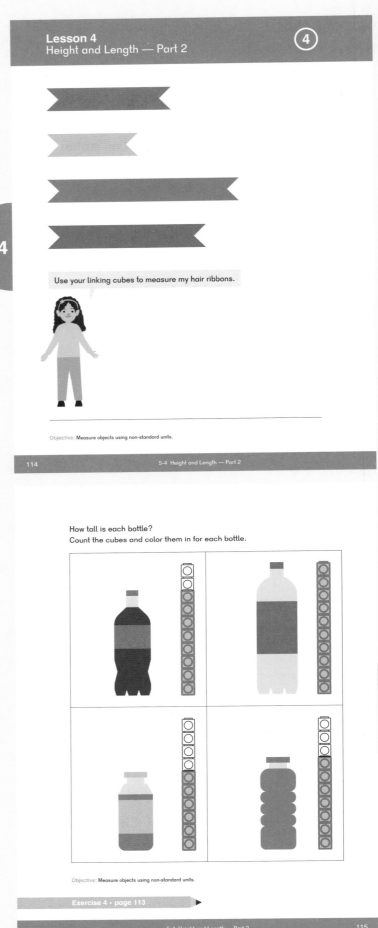

▲ Measure It

Materials: Linking cubes, objects to use as non-standard items to measure length such as craft sticks or unsharpened pencils, art paper, or paper clips

Provide students with linking cubes, craft sticks, and other objects that can be used as non-standard units for measuring, and some classroom objects (crayon, scissors, book, etc.) to measure. Have students choose one object to measure, using all of the different non-standard units.

Have students draw pictures of the item measured and the non-standard unit used to record their findings.

For example, a student may draw a picture of a book with craft sticks alongside to show how many craft sticks long the book measured.

▲ Estimation Game: Length

Materials: Linking cubes

Students play in pairs. Player 1 finds a classroom object to measure with linking cubes. Both players record their estimates of the object's length on a whiteboard. Player 1 measures the object with the cubes and the player whose estimate is closer to the actual length, without going over, scores a point.

For example, Player 1 estimates the length to be 6 cubes. Player 2 estimates the length to be 4 cubes. The actual length is 5 cubes, so Player 2 scores a point for being the closest without going over. If students have the same estimate, both can score a point if the estimate is within 1 cube.

▲ Roll and Draw

Materials: Die (BLM), linking cubes

Provide students with a Die (BLM) and linking cubes. Player 1 rolls the die and draws a line, trying to draw it as long as a linking cube train in which the number of cubes matches the number rolled. For example, Player 1 rolls a 4, and tries to draw a line that will measure 4 cubes long (without using cubes). Player 2 then builds a linking cube train that is 4 cubes long to compare.

If the line is shorter or longer, the line drawer scores one point. If the line is exactly the same length, 2 points are scored. Players switch roles and play continues.

 Exercise 4 • page 113

Extend

★ How Many Pieces of Ribbon?

Materials: Ribbons used in lesson

Have students estimate how far it is from one object in the classroom to another, using their piece of ribbon to measure. Then have them check their answers.

Lesson 5 Weight — Part 1

Objective

• Compare weight.

Lesson Materials

• *Balancing Act* by Ellen Stoll Walsh

Explore

Have students select a classroom object. With partners, ask students to compare the weight of the two objects, using the phrases "lighter than" and "heavier than." Have students partner up with a different student and repeat. Ask students to explain how they know that one object is heavier or lighter than the other.

Learn

Look at page 117 and discuss the illustration. Ask students to explain how they know that one object is heavier than the other without holding it.

Whole Group Activities

▲ I Spy

Play a game of **I Spy** using weight as a descriptor. For example, to describe scissors, a student might say, "I spy, with my little eye, something heavier than a pencil and lighter than a book."

▲ Reading Time

Materials: *Balancing Act* by Ellen Stoll Walsh

Read *Balancing Act* and have students act out the book as it is read aloud.

Small Group Activities

▲ Textbook Page 117

Lesson 5
Weight — Part 1   (5)

Look and talk.
Which thing is heavier?

Objective: Compare weight.

116   5-5 Weight — Part 1

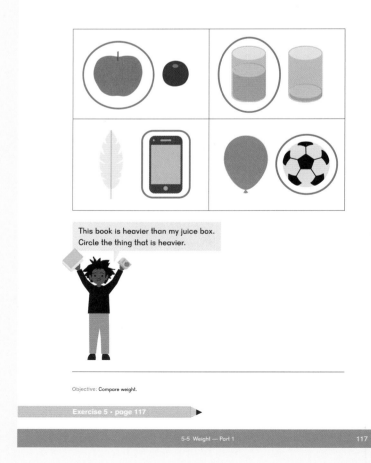

This book is heavier than my juice box.
Circle the thing that is heavier.

Objective: Compare weight.

Exercise 5 · page 117

5-5 Weight — Part 1   117

▲ Which is Heavier?

Materials: Comparing Weight (BLM)

Allow students to explore the classroom, comparing the weights of objects. For example, students could compare the weight of a book to the weight of a pencil. Students can record their findings on Comparing Weight (BLM).

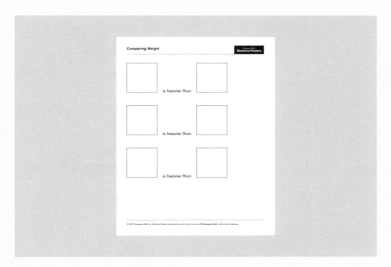

Exercise 5 • page 117 ▶

Extend

★ Airplanes

Materials: Paper clips, paper

Have students make paper airplanes and review height and length by seeing whose plane flies the greatest length from a starting point. Add weight in the form of paper clips to see how that affects the length or height that the airplane can fly.

Lesson 6 Weight — Part 2

Objective

• Compare weight.

Lesson Materials

• Balance scales, enough for partner or small group work
• *Who Sank the Boat?* by Pamela Allen

Explore

Show students a balance scale. Put a single object on either side and observe what happens. Have students work in partners or small groups to explore what happens to the scale when objects from the classroom are placed on either side.

Learn

Have students discuss the objects they weighed and compare which were heavier or lighter.

Look at page 118 and talk about the illustration.

Whole Group Activity

▲ Reading Time

Materials: *Who Sank the Boat?* by Pamela Allen, tin foil, small tub of water

Read *Who Sank the Boat?* aloud to the students. Have them fold a paper or tin foil boat and float it in a tub of water. Have them add objects and reenact the story or create a new story. Was it the cube that sank the boat?

Small Group Activity

▲ Which is Heavier? — Partners

Materials: Objects of varying weight, balance scales

Turn the activity from the previous lesson into a partner game. Have each student choose an object, and the student with the heaviest object in each pair scores a point.

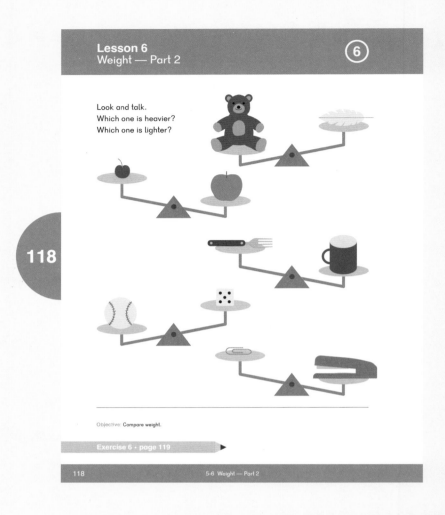

118

Lesson 6
Weight — Part 2   (6)

Look and talk.
Which one is heavier?
Which one is lighter?

Objective: Compare weight.

Exercise 6 • page 119

118   5-6 Weight — Part 2

◀ **Exercise 6 • page 119**

Extend

★ Which is Heaviest?

Materials: Objects of varying weight, balance scales

Students compare the weights of up to five objects using a balance scale and order the objects from lightest to heaviest.

Lesson 7 Weight — Part 3

Objective

- Measure weight with non-standard units.

Lesson Materials

- Balance scales, enough for partner or small group work
- Linking cubes
- Small objects

Explore

Have students recall when they measured how long an object was with linking cubes. Ask if they could also use the linking cubes to compare how much objects weigh.

Learn

Show students the balance scale. On one side, place an object that weighs less than 10 linking cubes. Place linking cubes on the other side of the scale, counting each one as it is set down, to balance the scale and find how many linking cubes the object weighs. Repeat with another object.

Ask, "What do you notice about the weight of the object and the number of cubes needed?"

Small Group Activity

▲ **Textbook** Pages 119 – 120

Ask students, "How do you know which is heavier, the toy car or the pencil?" They may respond, "The toy car is as heavy as 5 cubes, the pencil is as heavy as 3 cubes. The toy car needs more cubes than the pencil, so the toy car is heavier."

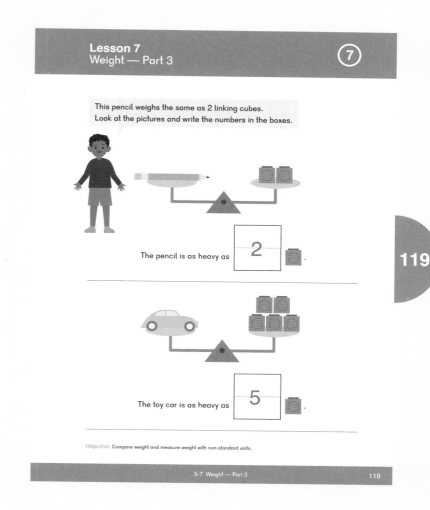

This pencil weighs the same as 2 linking cubes.
Look at the pictures and write the numbers in the boxes.

The pencil is as heavy as ⟶ 2 .

The toy car is as heavy as 5 .

Objective: Compare weight and measure weight with non-standard units.

5-7 Weight — Part 3   119

119

Extend

★ Box Weight

Materials: Boxes or containers of food of varying weights

Have students compare the weights of various boxes or containers of food. Students can discuss why different packages of cereal come in big boxes, but are lighter than smaller juice boxes.

120

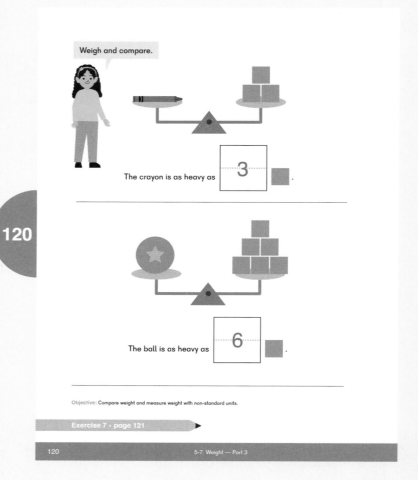

Lesson 8 Capacity — Part 1

Objective

- Compare capacity.

Lesson Materials

- Pail or bucket
- Cup

Show students a pail and a cup and ask them, "If you wanted to fill up a bathtub with water, which container would you use, the cup or the pail?"

Students should quickly come up with an answer such as, "The pail holds a lot more than the cup, so it is faster to use a pail."

Show students containers and discuss how the containers are used in real life. Have them try to put them in order from what would hold the least to the greatest amount (smallest to largest capacity).

Learn

Have students discuss what is happening with the water being poured in the illustration on textbook page 121 by comparing which containers hold more or less water than others.

- Mei's can holds less than her bowl.
- Alex's cup holds less than his pot.
- It's hard to see if Dion's milk jug holds more or less than his bowl.
- Sofia's bottle holds more than her cup.
- Emma's cup holds less than her pail.

Whole Group Activity

▲ I Spy

Play a game of **I Spy** using capacity as a descriptor. For example, a student might say, "I spy with my little eye something that holds more than a water bottle and less than a bathtub." (Playground ball bucket)

Look and talk.
Which holds more?

Objective: Compare capacity.

Small Group Activities

▲ **Textbook** Page 122

▲ **Which Holds More?**

Materials: Containers of various sizes (bowls, cups, pails, and bottles), materials to measure capacity (rice, beans, sand, seeds, or water)

Provide students with containers and different items to fill the containers. Rice, beans, sand, seeds, or water could be at different centers in the room.

▲ **Reading Time**

Materials: *Goldilocks and the Three Bears*, art paper, markers or crayons

Read *Goldilocks and the Three Bears* aloud and discuss. Review other measurements in the story.

- Why does Papa Bear get the biggest bowl and Baby Bear get the smallest bowl?
- Is Mama Bear's bed longer or shorter than Papa Bear's Bed? What about the chairs?

Students can draw a picture of Papa Bear, Mama Bear, Baby Bear, and their 3 bowls.

Exercise 8 • page 123

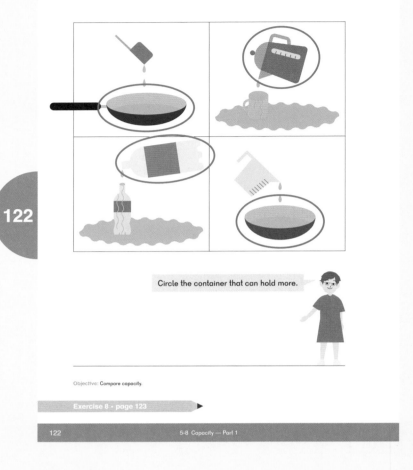

122

Circle the container that can hold more.

Objective: Compare capacity.

Exercise 8 • page 123

Extend

★ **Which Has the Most?**

Materials: 3 clear plastic jugs of varying size, water, food coloring

Prepare for this activity by filling 3 different clear plastic jugs of varying size with different-colored water, filling each to a different level. Save the other jugs for Lesson 10.

Have students vote by color which bottle has the most water.

At the end of this chapter, have students check their answers by providing similar empty containers and having students use the strategies they will learn in the next lesson.

Lesson 9 Capacity — Part 2

Objective

- Measure capacity with non-standard units.

Lesson Materials

- 2 cups that obviously hold different amounts
- Sets of 2 similar plastic cups, one slightly wider and shorter than the other
- Small drinking cups
- A variety of containers such as bowls, pails, and pots
- Rice, beans, or water

123

I wonder how many cups of sand this pail holds.

Objective: Measure capacity with non-standard units.

Explore

Show students two cups that obviously hold different amounts of water. Ask which holds more. Next, show students two similar sized cups and ask the same question. The answer should not be obvious.

After discussion, provide groups of three or four students two of the similar cups and a tub of either water or beans to find an answer. Put out a variety of other containers, including some small drinking cups, and tell students they can use whatever they want to try to figure the problem out.

Students may use different strategies:

- Fill up one cup, and then pour it in the other to see if there's room left or if it overflows.
- Use the small drinking cups to see how many it takes to fill each cup and compare the answers.
- Fill up both cups and then pour them into two identical containers to compare.

After exploring, have students share which cup held more and then demonstrate what they did. Have them share what worked and what didn't.

Learn

Have students discuss what Dion and Sofia are doing with their cups on page 123.

Small Group Activities

▲ **Textbook** Page 124

▲ **How Much Does it Hold?**

Materials: A variety of containers, materials to measure (rice, beans, or water), small drinking cups

Provide students with containers, including small square sandwich containers and other non-standard shaped containers. Have students count how many of the small drinking cups of rice, beans, or water it takes to fill the larger container.

▲ **Estimation Game: Capacity**

Materials: A variety of containers, materials to measure (rice, beans, or water), small drinking cups

With a partner, each student chooses a container to fill. Both students estimate and then check to see how many small drinking cups will fill the container.

Take it Outside

▲ **Build a Sand Castle**

Materials: Wet sand, cups of different sizes

Provide students with wet sand and cups of different sizes. Allow them to work in groups to build a castle. Review height by seeing which group built the tallest castle in the allotted time.

Exercise 9 • page 125

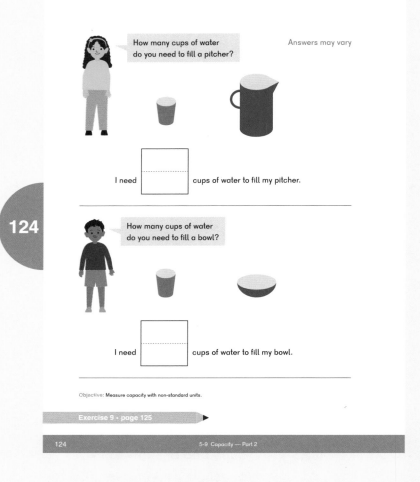

How many cups of water do you need to fill a pitcher?

Answers may vary

I need [ ] cups of water to fill my pitcher.

How many cups of water do you need to fill a bowl?

I need [ ] cups of water to fill my bowl.

Objective: Measure capacity with non-standard units.

Exercise 9 • page 125

124   5-9 Capacity — Part 2

Extend

★ **Leftovers**

Materials: Rice, cups, bowls of varying sizes

Provide students with problems involving counting. Have them use rice and multiple same-sized cups to measure. For example, Dion's bucket is full. It holds 5 cups of rice. Emma's bucket can hold 8 cups of rice. If Dion pours his rice into Emma's bucket, how many more cups of rice will fill Emma's bucket?

Teacher's Guide KA Chapter 5   © 2017 Singapore Math Inc.

Lesson 10 Practice

Objective

- Practice concepts from the chapter.

Students can complete the textbook pages and workbook pages as practice and/or as assessment.

Use center **Activities** from the chapter for additional review and practice.

▲ Paper Chains

Materials: Construction paper cut into strips

Provide students with strips of construction paper and have them create paper chains by stapling strips into interconnecting loops. After creating their chains, have them order the chains from shortest to longest. Have each student make a chain and display them in the classroom. The chains can be used in **Chapter 6**, **Lesson 3**: **More and Less** to compare quantities to 10.

▲ Mother May I?

Have students line up in a row. Stand about 20 feet from them. Have them estimate how many steps it would take for them to reach you. Play **Mother May I?** A student may ask, "Mother may I take 15 steps?" That student takes 15 steps. Another students then asks, adjusting her question based on what she observed.

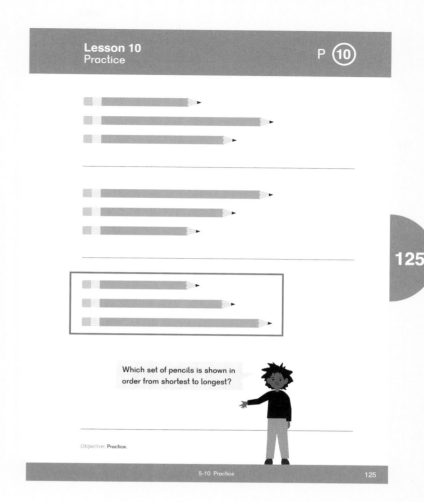

Which set of pencils is shown in order from shortest to longest?

Objective: Practice.

5-10 Practice   125

125

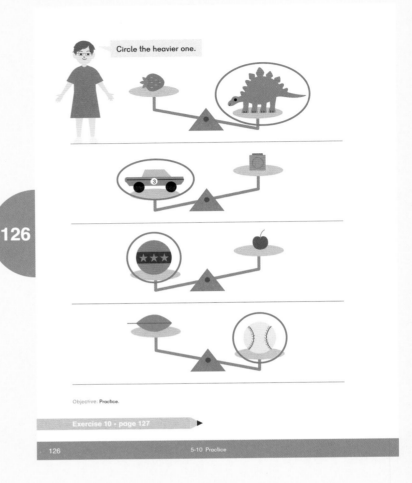

Extend

★ Which Has the Most?

Materials: Empty cups of equivalent size, containers with colored water from **Lesson 8: Capacity — Part 1**

Have students check their answer by providing similar empty containers and having students use the strategies they have learned by either direct or indirect comparison to see which jug has the most water.

★ Balloon Weight

Materials: Two identical balloons, water

Fill one of two identical balloons completely with air. Put a small amount of water in the second balloon.

Have students predict which balloon will be heavier and share their reasoning. They can then weigh the two balloons to check their answer.

Ask them to draw a conclusion from the measurement. Students may notice that air must weigh less than water.

Chapter 5 Compare Height, Length, Weight, and Capacity

Exercise 1

Circle the tallest thing.
Cross out the shortest thing.

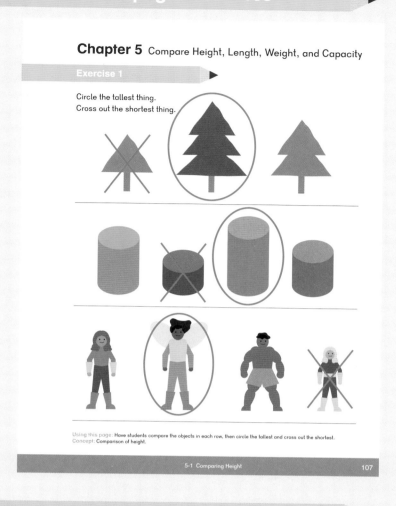

Using this page: Have students compare the objects in each row, then circle the tallest and cross out the shortest.
Concept: Comparison of height.

5-1 Comparing Height   107

Order the giraffes from tall to tallest.

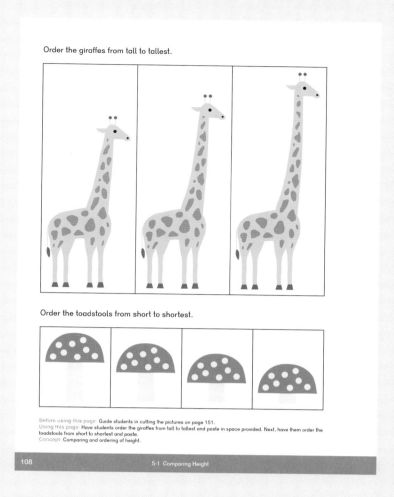

Order the toadstools from short to shortest.

Before using this page: Guide students in cutting the pictures on page 151.
Using this page: Have students order the giraffes from tall to tallest and paste in space provided. Next, have them order the toadstools from short to shortest and paste.
Concept: Comparing and ordering of height.

108   5-1 Comparing Height

Exercise 2

Color the longest thing yellow.
Color the shortest thing pink.

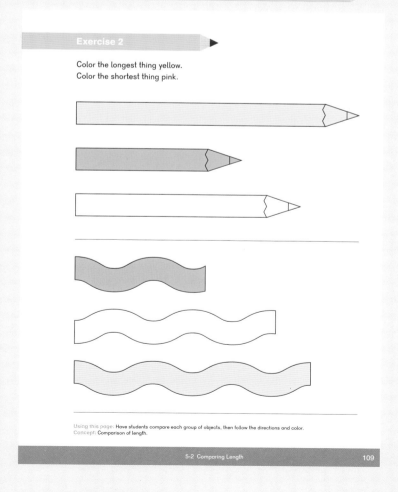

Using this page: Have students compare each group of objects, then follow the directions and color.
Concept: Comparison of length.

5-2 Comparing Length   109

Order the rulers from long to longest.

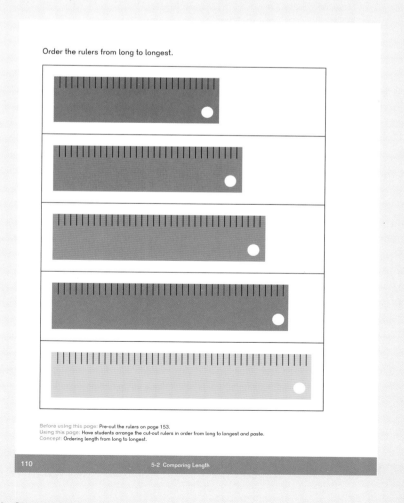

Before using this page: Pre-cut the rulers on page 153.
Using this page: Have students arrange the cut-out rulers in order from long to longest and paste.
Concept: Ordering length from long to longest.

110   5-2 Comparing Length

   Teacher's Guide KA Chapter 5

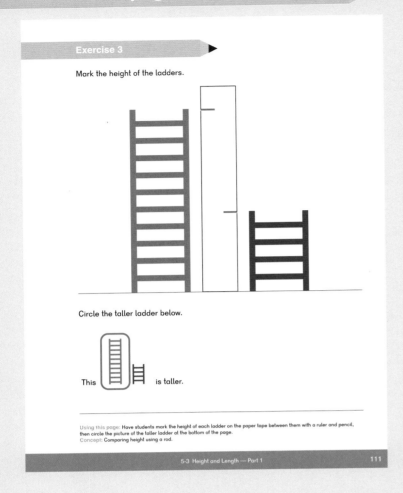

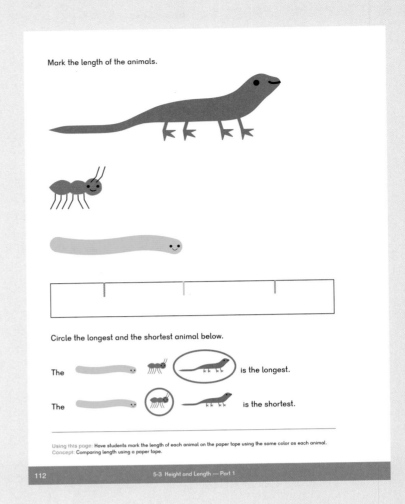

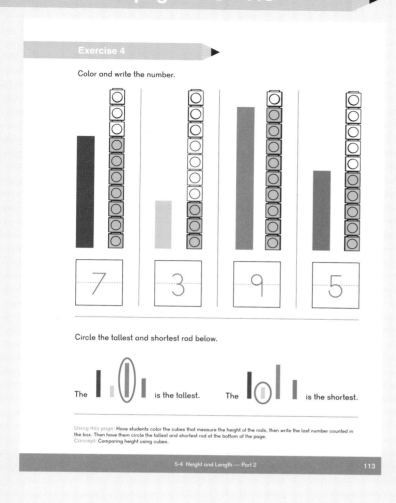

Exercise 4

Color and write the number.

Circle the tallest and shortest rod below.

The ___ is the tallest.   The ___ is the shortest.

Using this page: Have students color the cubes that measure the height of the rods, then write the last number counted in the box. Then have them circle the tallest and shortest rod at the bottom of the page.
Concept: Comparing height using cubes.

5-4 Height and Length — Part 2   113

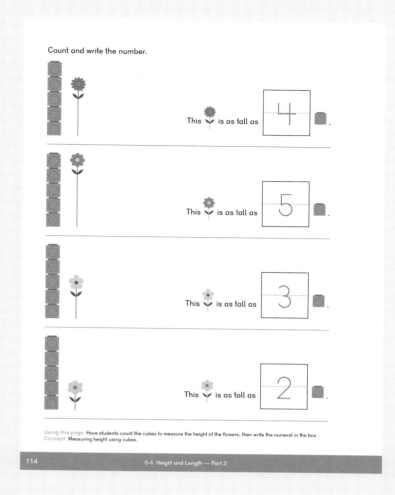

Count and write the number.

This 🌻 is as tall as 4 ▢.

This 🌻 is as tall as 5 ▢.

This 🌸 is as tall as 3 ▢.

This 🌸 is as tall as 2 ▢.

Using this page: Have students count the cubes to measure the height of the flowers, then write the numeral in the box.
Concept: Measuring height using cubes.

114   5-4 Height and Length — Part 2

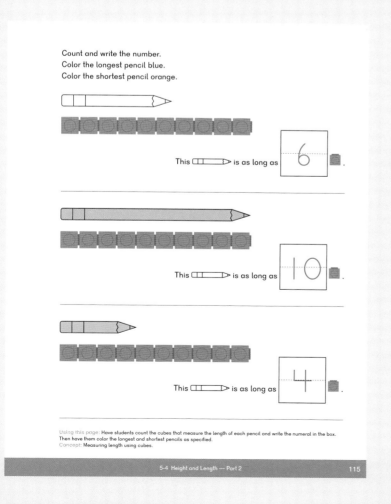

Count and write the number.
Color the longest pencil blue.
Color the shortest pencil orange.

This ✏ is as long as 6 ▢.

This ✏ is as long as 10 ▢.

This ✏ is as long as 4 ▢.

Using this page: Have students count the cubes that measure the length of each pencil and write the numeral in the box.
Then have them color the longest and shortest pencils as specified.
Concept: Measuring length using cubes.

5-4 Height and Length — Part 2   115

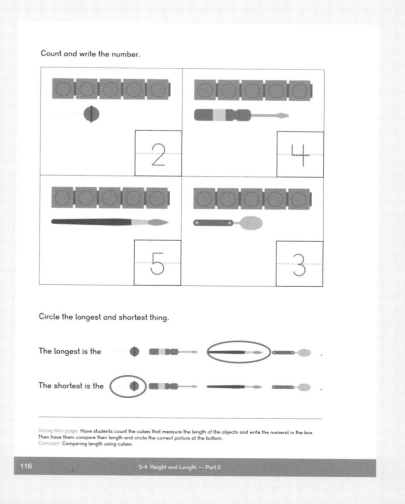

Count and write the number.

2   4

5   3

Circle the longest and shortest thing.

The longest is the ___

The shortest is the ___

Using this page: Have students count the cubes that measure the length of the objects and write the numeral in the box.
Then have them compare their length and circle the correct picture at the bottom.
Concept: Comparing length using cubes.

116   5-4 Height and Length — Part 2

© 2017 Singapore Math Inc.   Teacher's Guide KA Chapter 5   173

Exercise 5 • pages 117 – 118

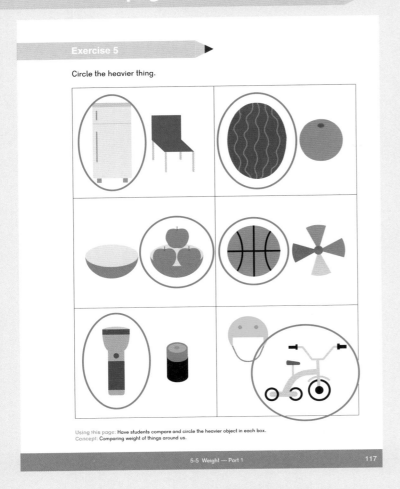

Exercise 5

Circle the heavier thing.

Using this page: Have students compare and circle the heavier object in each box.
Concept: Comparing weight of things around us.

5-5 Weight — Part 1   117

Circle the lighter one.

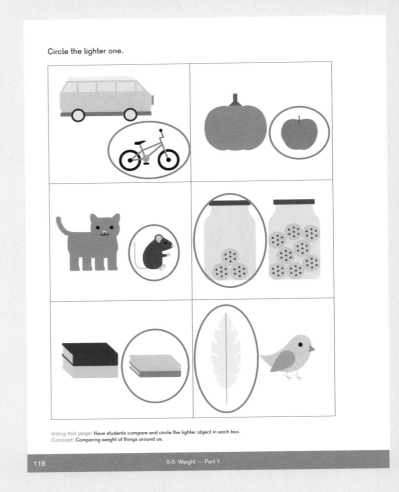

Using this page: Have students compare and circle the lighter object in each box.
Concept: Comparing weight of things around us.

118   5-5 Weight — Part 1

Exercise 6 • pages 119 – 120

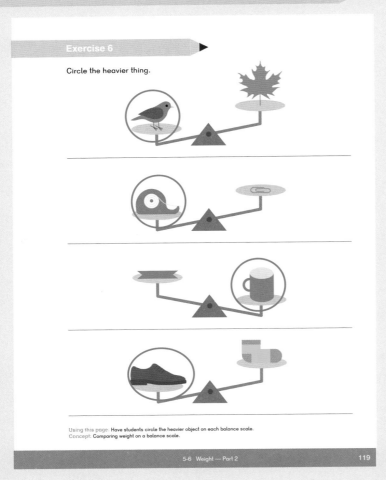

Exercise 6

Circle the heavier thing.

Using this page: Have students circle the heavier object on each balance scale.
Concept: Comparing weight on a balance scale.

5-6 Weight — Part 2   119

Circle the lighter thing.

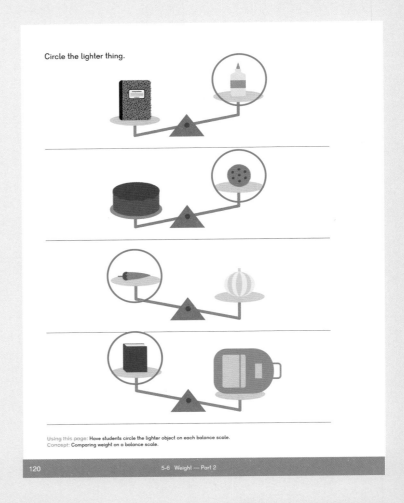

Using this page: Have students circle the lighter object on each balance scale.
Concept: Comparing weight on a balance scale.

120   5-6 Weight — Part 2

Exercise 7

Count and write the number.

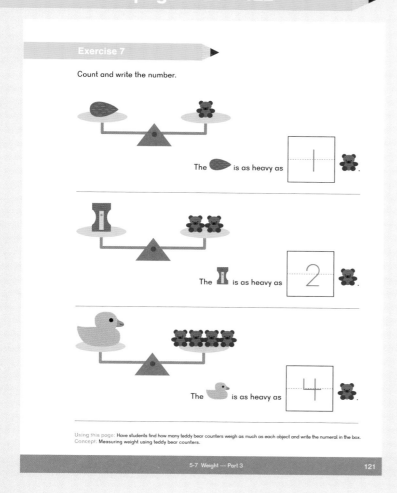

The 🌰 is as heavy as ⬜1⬜ 🧸.

The ⌛ is as heavy as ⬜2⬜ 🧸.

The 🦆 is as heavy as ⬜4⬜ 🧸.

Using this page: Have students find how many teddy bear counters weigh as much as each object and write the numeral in the box.
Concept: Measuring weight using teddy bear counters.

5-7 Weight — Part 3   121

Count and write the number.

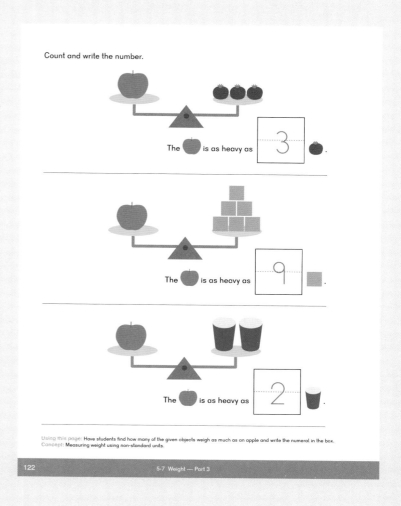

The 🍎 is as heavy as ⬜3⬜ 🫐.

The 🍎 is as heavy as ⬜9⬜ ⬜.

The 🍎 is as heavy as ⬜2⬜ 🥤.

Using this page: Have students find how many of the given objects weigh as much as an apple and write the numeral in the box.
Concept: Measuring weight using non-standard units.

122   5-7 Weight — Part 3

Exercise 8

Circle the container that can hold more.

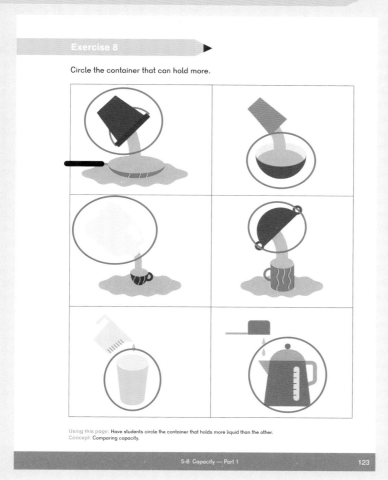

Using this page: Have students circle the container that holds more liquid than the other.
Concept: Comparing capacity.

5-8 Capacity — Part 1   123

Circle the container that holds less.

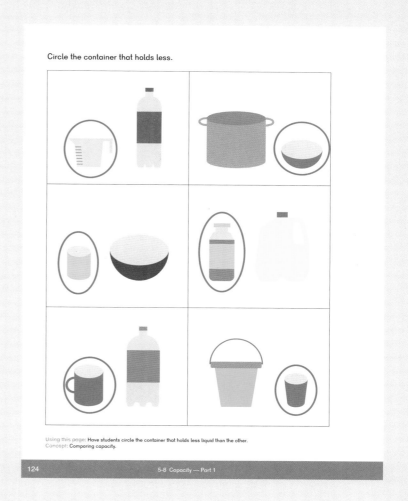

Using this page: Have students circle the container that holds less liquid than the other.
Concept: Comparing capacity.

124   5-8 Capacity — Part 1

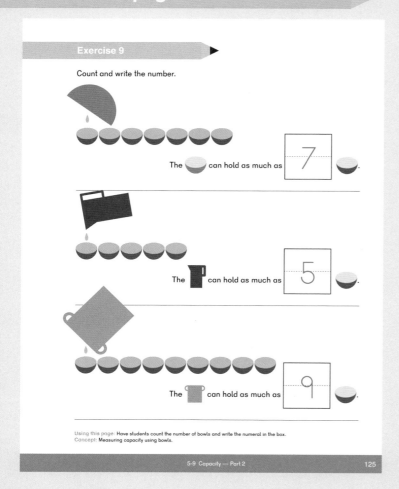

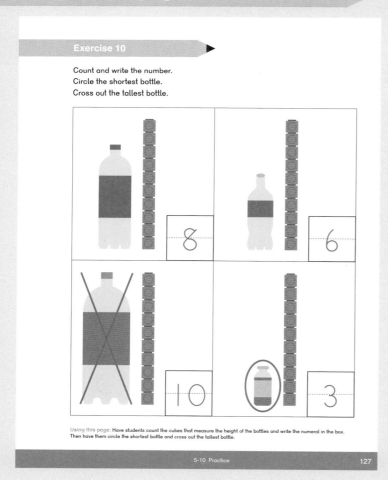

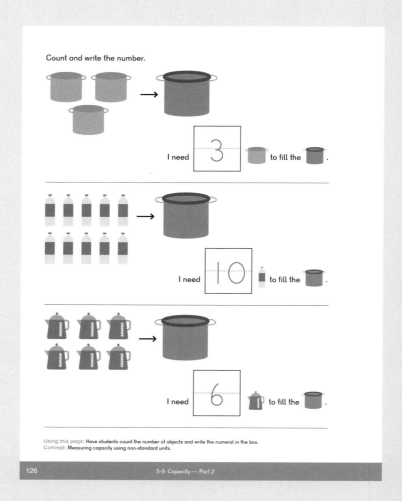

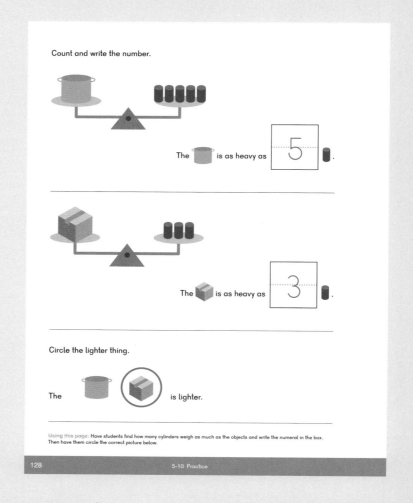

Teacher's Guide KA Chapter 5   © 2017 Singapore Math Inc.

Suggested number of class periods: 5 – 6

Lesson	Page	Resources		Objectives
Chapter Opener	p. 181	TB:	p. 127	
1 Same and More	p. 182	TB: WB:	p. 128 p. 129	Recognize equal groups. Identify a group that has more objects than another group.
2 More and Fewer	p. 184	TB: WB:	p. 130 p. 133	Identify a group that has more and a group that has fewer objects than another group.
3 More and Less	p. 187	TB: WB:	p. 133 p. 137	Compare sets of objects using "more" and "less."
4 Practice — Part 1	p. 190	TB: WB:	p. 137 p. 141	Practice concepts from the chapter.
5 Practice — Part 2	p. 192	TB: WB:	p. 141 p. 143	Practice concepts from the chapter.
Workbook Solutions	p. 193			

This chapter teaches mathematical comparison. While students have begun comparing objects by attribute (larger or smaller, taller or shorter, etc.), mathematical comparison involves comparing objects by quantity.

In Kindergarten, students begin to understand the abstract order of numbers by relating the quantities in groups.

Students are taught first to recognize when two groups have equal quantities, and then recognize the relationship between two groups when one has more than another, and therefore, one has less than another.

Initially, students are asked to make direct comparisons of small sets of objects by counting. Students then begin to visually identify sets with more and fewer objects. Lastly, they are asked to realize that 6 of an item is less than 7 of an item, no matter the item. Ultimately, students should move from a concrete comparison of the quantities to an abstract comparison of the numbers without seeing the quantity.

The main phrases for comparison are "more than," "greater than," "fewer than," and "less than." Traditionally, "fewer than" is used for items that can be made plural: "I have fewer ducks than elephants," while "less than" is used for items that remain singular (representing an aggregate): "I have less water than sand." For this reason, numbers by themselves are compared using "less than."

"More than" is a phrase often used for both singular and plural items: "I have more fish than birds." The custom is to use "greater than" when comparing numbers. 7 is greater than 5, 5 is less than 7, however, 2 more than 5 is 7.

Teachers should model proper use of these phrases, but students are not expected to use these terms themselves in **Dimensions Math® Kindergarten**.

Avoid using the terms "bigger" and "smaller" as that can lead to confusion. For example, which is bigger:

3 or 7?

Students often struggle conceptually with "less than" and "fewer than" even when they grasp "more than." Help students by mentioning both relationships as often as possible. For example, if I have more apples than oranges, then I have fewer oranges than apples.

While this chapter is short, it contains many important and oftentimes difficult concepts for students. You may choose to take more than one day to teach a lesson or focus on the activities. Students will continue to practice these concepts in **Dimensions Math® Kindergarten B**, and will have another opportunity to master comparison in **Dimensions Math® 1A**.

Storybooks that go along with content in this chapter can be added as time allows or for additional exploration. Storybook suggestions are listed on the following page.

Materials

- Linking cubes
- Beans, beads, bear (or other animal) counters
- Dice
- Playing cards

Blackline Masters

- Ten-frame Cards
- Number Cards
- Comparing Numbers
- Single-line Ten-frames

Storybooks

- *More, Fewer, Less* by Tana Hoban
- *Just Enough Carrots* by Stuart J. Murphy
- *Peg + Cat: The Race Car Problem* by Jennifer Oxley and Billy Aronson

Letters Home

- Chapter 6 Letter

Notes

Teacher's Guide KA Chapter 6

© 2017 Singapore Math Inc.

Materials

- *More, Fewer, Less* by Tana Hoban

This **Chapter Opener** is designed to introduce students to comparing numbers.

Have students discuss the objects in the picture on page 127.

Possible student observations:

- Each child has a pail.
- There are 3 pets and 2 children.
- There are 2 flowers and 2 children.
- There are 2 biscuits and 2 children (or 2 biscuits and 2 dogs).

They may use the terms "more than" or "less than" already, as in:

- There are more pets than children.
- There are less (fewer) biscuits than pets.

Read and discuss the pictures in the book *More, Fewer, Less* by Tana Hoban.

Take a "More, Fewer, Same" adventure walk around the classroom. In the classroom:

- Are there more boys than girls?
- Are there fewer boys than girls?
- Is there the same number?
- Are there more desks than students?

In the kitchen:

- Are there more cups, bowls, or plates on the table?
- Are there more apples than bananas?
- Are there more bowls on the table or in the cupboard?

Chapter 6

Comparing Numbers Within 10

Look and talk.
Is there a pail for each child?
Is there a child for each pet?

127

Objectives

- Recognize equal groups.
- Identify a group that has more objects than another group.

Lesson Materials

- Bags of 2 colors of counters with an equal number of each color in each bag, 1 per pair of students (less than 10 of each color)
- Single-line Ten-frames (BLM)

Explore

Give each pair of students a bag with an equal number of two different colors of counters. Have students count the counters. Ask, "Are there more or the same number of each color?"

For groups that count quickly and say, "There are 5 in each group, so they are the same," ask them to show or prove to you that they are the same. Encourage students to organize their counters in a way that shows that they are the same.

Learn

Call 5 students to the front of the class. Pass out 4 books to the students. Ask:

- What do you notice about the number of students and the number of books?
- Are there the same of each, or is there more of one or the other?

Have students show with counters and Single-line Ten-frames (BLM). Repeat the activity with 6 students and 6 books. Ask the class what they notice. How can they show there are the same number of students and books?

Discuss page 128.

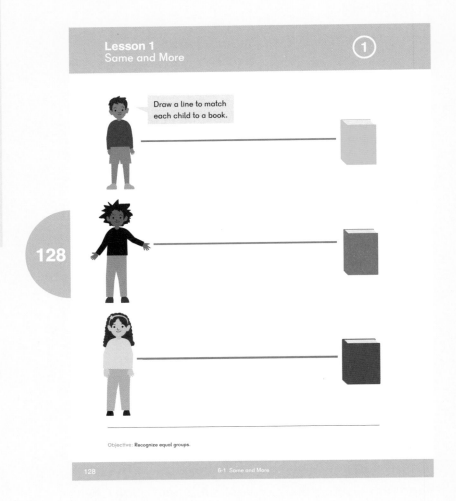

Objective: Recognize equal groups.

128   6-1 Same and More

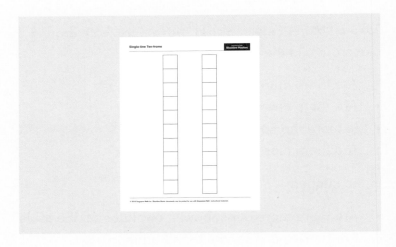

Whole Group Activity

▲ Musical Chairs

Materials: Chairs — 1 less than the number of students playing to start, music

Play musical chairs, emphasizing that in each round there will be more students than chairs.

Small Group Activities

▲ **Textbook** Pages 128 – 129

▲ **Domino Match**

Materials: Dominoes

Set up an array of up to double-five dominoes. Have students play in small groups and take turns to find a match. A match is two dominoes showing the same number of pips.

▲ **Domino Number Match**

Materials: Number Cards (BLM) 1 to 5, Dot Cards (BLM) 1 to 5, Ten-frame Cards (BLM) 1 to 5, Dominoes

Have students match a domino, Dot Card (BLM), or Ten-frame Card (BLM) to a Number Card (BLM).

▲ **Match**

Materials: Number Cards (BLM) 1 to 5, Dot Cards (BLM) 1 to 5, Ten-frame Cards (BLM) 1 to 5

Students arrange the cards, faceup, in a grid.
Students take turns finding two cards that go together.

★ **Memory**

Materials: Number Cards (BLM) 1 to 5, Dot Cards (BLM) 1 to 5, Ten-frame Cards (BLM) 1 to 5

Students arrange the cards, facedown, in a grid.
Students take turns finding two cards that go together.

Exercise 1 • page 129 ▶

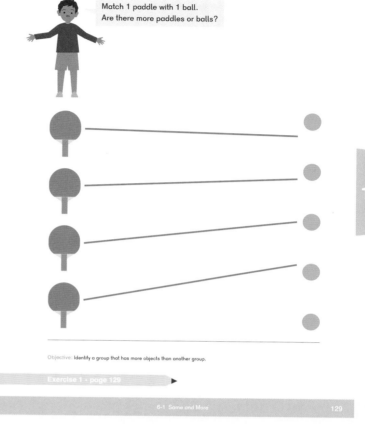

Match 1 paddle with 1 ball.
Are there more paddles or balls?

Objective: Identify a group that has more objects than another group.

Exercise 1 • page 129 ▶

6-1 Same and More   129

Extend

★ **More or Fewer Airball**

Materials: Craft sticks, paper plates, small balloons

Have students make ping-pong-style paddles with craft sticks and paper plates. Use small balloons for balls. Bat the balloons around keeping them from hitting the floor.

Have students use the terms "same," "more," and, "fewer" to describe the game. Examples:

- We have 2 hands and 1 paddle each, so there are more hands than paddles.
- There are fewer balloons than paddles.

Add some balloons and ask students to describe the game again.

Objective

- Identify a group that has more and a group that has fewer objects than another group.

Lesson Materials

- Sets of 2 colors of counters or linking cubes with a different amount of each color in bags for partners (Less than 10 of each color)
- Single-line Ten-frames (BLM)

Lesson 2
More and Fewer
②

Look and talk.
Which group has more?
Which group has fewer?

Objective: Identify a group which has more and a group which has fewer objects than another group.

130   6-2 More and Fewer

Explore

Repeat the prior lesson's activity, adding the term "fewer."

Give partners a bag with two different colors of counters. Ask students to place the counters on a Single-line Ten-frame (BLM) and have them compare the quantity of the counters. Ask:

- Are there more of one color than the other?
- Are there fewer of one color than the other?

Encourage students to organize their counters on a Single-line Ten-frame and say the comparison both ways:

- There are more red counters than blue counters.
- There are fewer blue counters than red counters.

Learn

Discuss page 130 and ask students what the textbook means by "group." Ask, "Are there fewer children than swings?"

Ask students to share what other items they can see that are "more" or "fewer."

Whole Group Activity

▲ Ten-frame Flash

Materials: Ten-frame Cards (BLM) 1 to 10

Hold up two Ten-frame Cards (BLM) showing different numbers. Have students point to the card showing "more" dots as quickly as possible. Repeat the activity with new cards and ask students to point to the card showing fewer dots.

Small Group Activities

▲ Textbook Pages 131 – 132

Emphasize finding the answer by matching up shirts and shorts one by one:

- 1 shirt and 1 short, 2 shirts and 2 shorts, 3 shirts and 3 shorts, 4 shirts and 4 shorts, 5 shirts but not 5 shorts.

▲ Roll and Compare

Materials: Blank Ten-frames (BLM), linking cubes (2 different colors) or two-color counters, modified die with 3 sides that match one color of counter and 3 sides that match the other color of counter

Students roll the die, take a counter of the color shown on the die, and place it in a square on the Blank Ten-frame (BLM). Continue rolling the die and placing a counter until the whole ten-frame is full (one counter per square). Have students sort and count how many there are of each color. Which color has the most, fewest, or same amount?

Circle the group that has more.

Objective: Identify a group which has more than another group.

6-2 More and Fewer   131

131

Circle the group that has fewer.

Objective: Identify a group that has fewer objects than another group.

Exercise 2 · page 133

132   6-2 More and Fewer

▲ Difference Game

Materials: Die, linking cubes

Player 1 rolls a die and builds a tower using the corresponding number of linking cubes. Player 2 rolls and builds. Players compare their towers side-by-side. The player with more snaps off the difference and keeps those as scoring cubes for the round, the rest are returned to the pile. If the numbers are the same, no points are scored. Play continues until time is up. The player with more cubes at the end wins.

Modify the game for "fewer" by having the winner be the player with the fewest cubes at the end of the game.

▲ Face-off

Materials: Ten-frame Cards (BLM) 0 to 10

Each pair of players will need several sets of Ten-frame Cards (BLM). Deal the cards equally between the players. Each player flips a card faceup. The player with more on her ten-frame card wins both cards. If the number of dots showing on both cards is the same, a face-off begins and each player flips another card and compares until one player has more. The player with more dots showing wins all the cards.

Modify the game for "fewer" so that the player with fewer counters wins the cards.

Exercise 2 • page 133 ▶

Extend

★ More or Fewer?

Materials: Ten-frame Cards (BLM) 0 to 10

Provide each player with a set of Ten-frame Cards (BLM) 0 to 10.

Player 1 selects a "secret" card from her hand and places it facedown. Player 2 tries to guess the number on the "secret" card by laying down a card from his hand, faceup. Player 1 then tells whether the value shown on the "secret" card is more or less than Player 2's faceup card.

Player 2 continues selecting and showing different cards until he finds the value of the "secret" card. Players then switch roles.

Teacher's Guide KA Chapter 6

Objective

- Compare sets of objects using "more" and "less."

Lesson Materials

- Index cards, 11 for each student or pair of students

Explore

Begin by having students practice writing the numerals 0 to 10 by making number cards for themselves (could be done in pairs). If students finish the task quickly, have them put the cards in order from 0 to 10, or play **What's Missing?** with a partner.

▲ What's Missing?

Materials: Number cards created in **Explore**

Partners play with one set of the number cards 0 to 10 they just created. Shuffle cards. Player 1 pulls one number from the deck as the "secret number," then gives the remaining deck of cards to Player 2. Player 2 orders the cards and finds the missing number.

Learn

With their cards, ask students to find and show a number that is greater than 5. Ask students to show you a number that is less than 5. Working with partners, have students show you two numbers that are greater than 6 and then two numbers that are less than 6. Repeat the activity choosing different numbers.

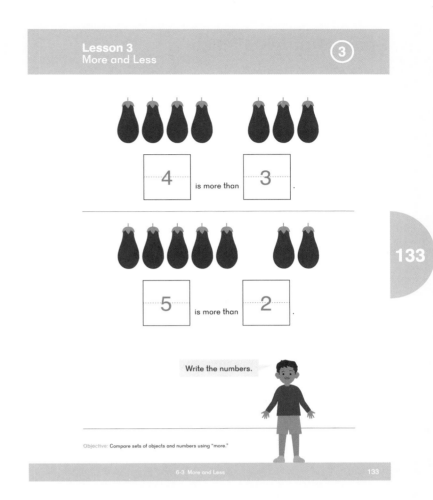

Small Group Activities

▲ **Textbook** Pages 133 – 135

Read and model the sentences: 4 eggplants are more than 3 eggplants, so 4 is more than 3.

▲ **Paper Chains**

Materials: Comparing Numbers (BLM), Paper chains from **Chapter 5**: **Lesson 10**

Have students count the links in the paper chains from **Chapter 5**: **Lesson 10**. Have students use Comparing Numbers (BLM) to say which classmate's chain has more or fewer links. For example, "Samitha's chain has 7 links, mine has 5. 7 is greater than 5 and 5 is less than 7."

▲ **Comparing Numbers**

Materials: Comparing Numbers (BLM), Number Cards (BLM) 0 to 10, counters

Provide students with Comparing Numbers (BLM), counters, and sets of Number Cards (BLM) 0 to 10. Students draw two cards and make the number with counters. For each pair of cards drawn, students complete a line on the sheet: "____ is less than ____." Players then switch roles.

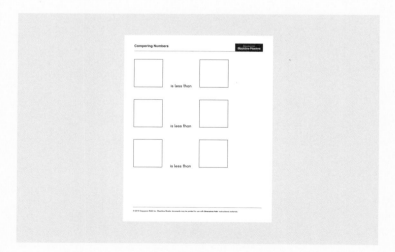

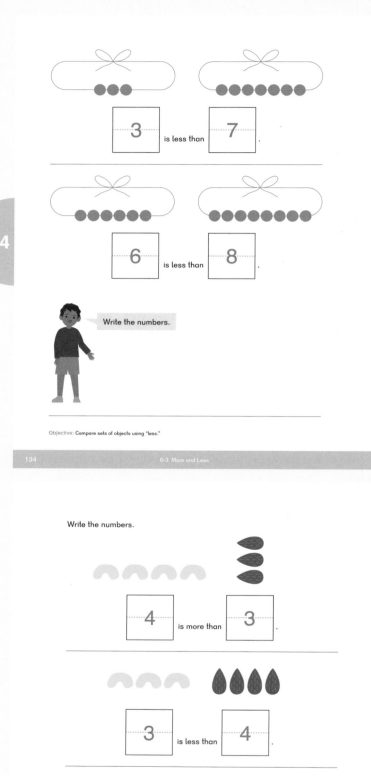

Page 136 is an extension. Complete one or more of the activities below before completing the page. Have students draw lines from the guitars to the friends to see how many more friends there are than guitars.

▲ How Many More Face-off

Materials: Ten-frame Cards (BLM) 0 to 10

Laminate sets of Ten-frame Cards (BLM) so that students can write on them with markers.

Have partners play **Face-off** from **Lesson 2** with laminated Ten-frame Cards (BLM). Players can use a dry erase marker to cross off the number that is the same, to see how many more dots are on the "greater" card.

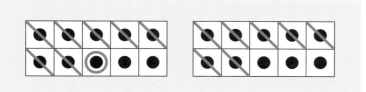

▲ Match Me

Materials: Art paper, pencils

Have students draw a picture of two groups of objects that go together. Tell students that there should be one group that has more than the other. Students trade pictures with each other and discuss which group has more, and how many more of the second group are needed.

For example, there are 6 frogs and 5 lily pads. We need 1 more lily pad so each frog has its own lily pad. Or 6 children and 4 bikes, we need 2 more bikes so each child can go on the bike ride.

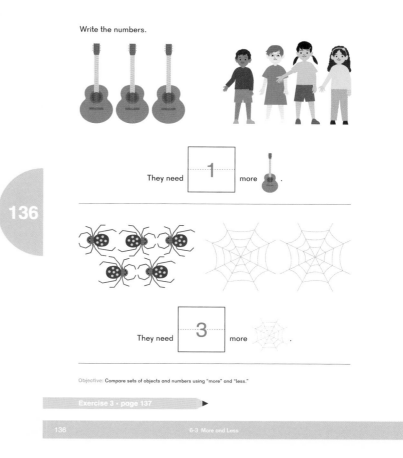

Write the numbers.

They need [1] more [guitar]

They need [3] more [web]

Objective: Compare sets of objects and numbers using "more" and "less."

Exercise 3 • page 137

136   6-3 More and Less

Extend

★ Rock Paper Scissors Math — More!

Partners bounce fists in their other hand while saying, "Rock, paper, scissors, math."

On the word "math," each player shoots out some fingers. The first player to say how many more fingers the player with more has showing is the winner.

Example: Player 1 shows 6 fingers, Player 2 shows 8 fingers. The first player to say, "8 is 2 more than 6," is the winner.

Exercise 3 • page 137

Lesson Materials

- Practice concepts from the chapter.

Practice lessons are designed for further practice and assessment as needed.

Students can complete the textbook pages and workbook pages as practice and/or as assessment.

Use **Activities** from the chapter for additional review and practice.

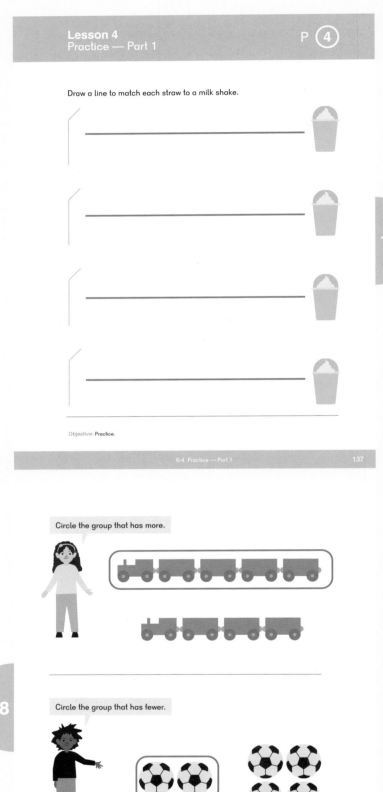

Lesson 4
Practice — Part 1   P 4

Draw a line to match each straw to a milk shake.

Objective: **Practice.**

6-4 Practice — Part 1   137

137

Circle the group that has more.

138

Circle the group that has fewer.

Objective: **Practice.**

138   6-4 Practice — Part 1

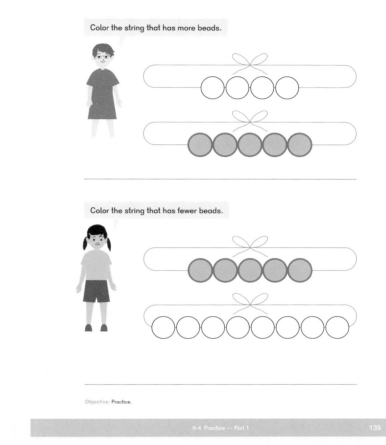

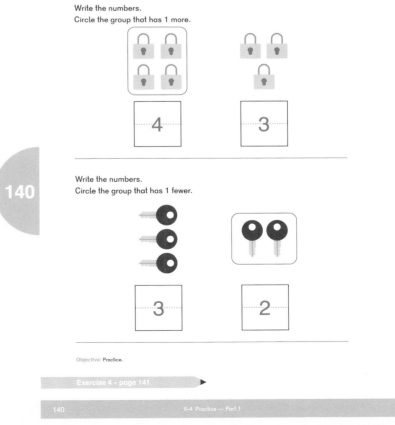

© 2017 Singapore Math Inc.   Teacher's Guide KA Chapter 6   191

Lesson 5 Practice — Part 2

Objective

- Practice concepts from the chapter.

Exercise 5 • page 143

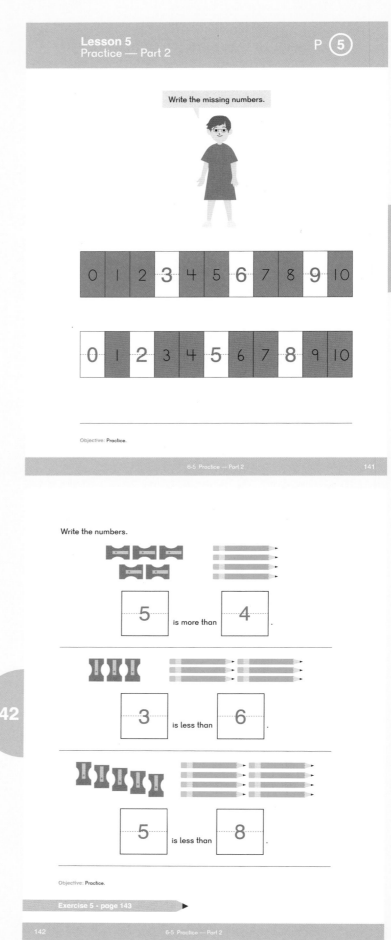

Chapter 6 Comparing Numbers Within 10

Circle the 2 groups that have the same number.

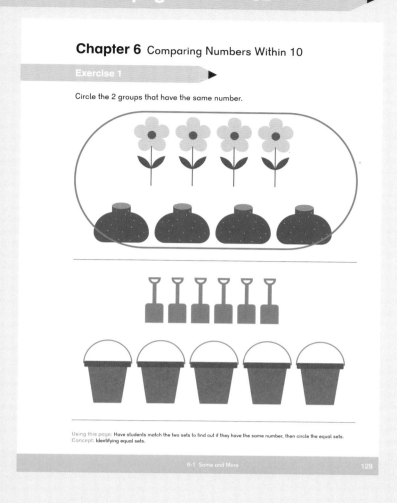

Using this page: Have students match the two sets to find out if they have the same number, then circle the equal sets.
Concept: Identifying equal sets.

6-1 Some and More   129

Color the 2 groups with the same number.

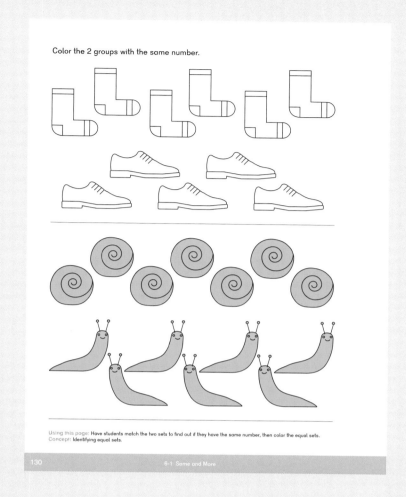

Using this page: Have students match the two sets to find out if they have the same number, then color the equal sets.
Concept: Identifying equal sets.

130   6-1 Some and More

Circle the group that has more.

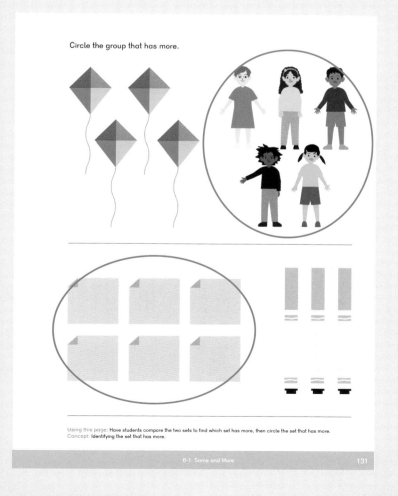

Using this page: Have students compare the two sets to find which set has more, then circle the set that has more.
Concept: Identifying the set that has more.

6-1 Some and More   131

Color the group that has more.

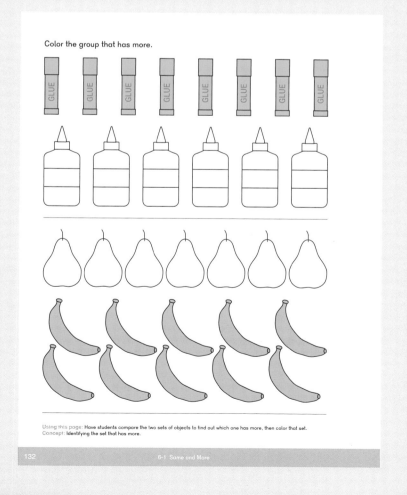

Using this page: Have students compare the two sets of objects to find out which one has more, then color that set.
Concept: Identifying the set that has more.

132   6-1 Some and More

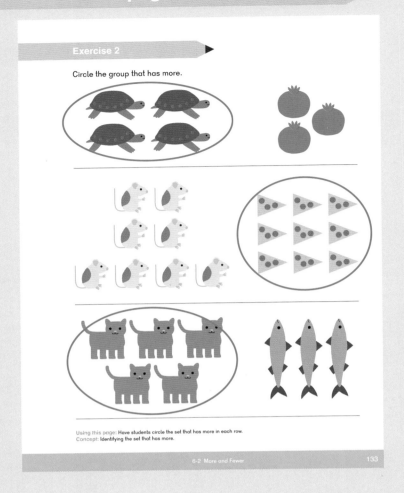

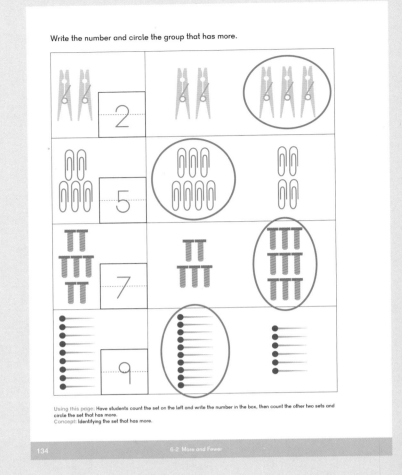

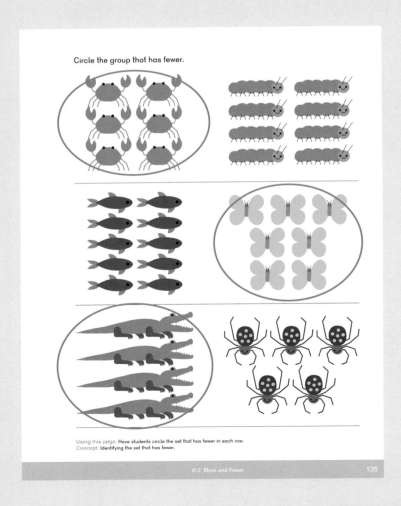

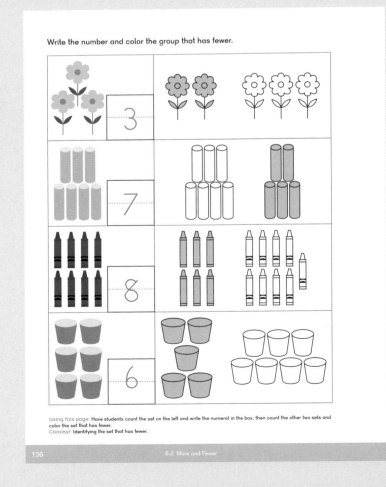

Exercise 3

Compare and write the number.

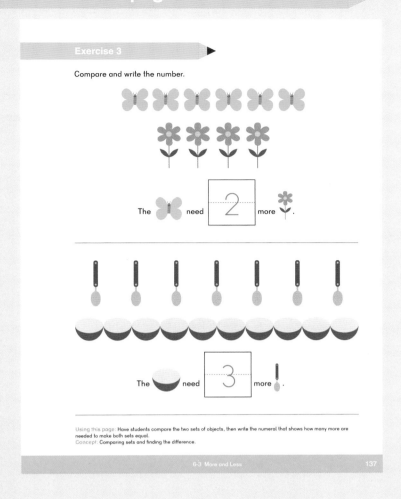

The 🦋 need **2** more 🌼.

The 🥣 need **3** more 🥄.

Using this page: Have students compare the two sets of objects, then write the numeral that shows how many more are needed to make both sets equal.
Concept: **Comparing sets and finding the difference.**

6-3 More and Less · 137

Compare and write the number.

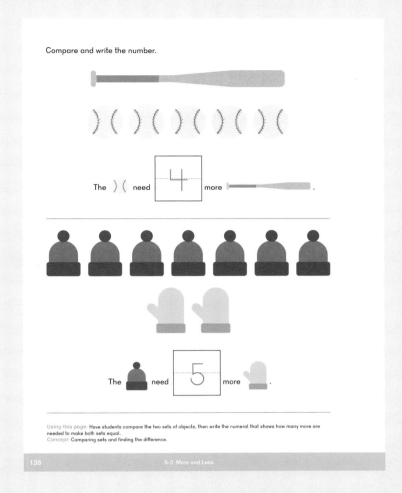

The)(need **4** more ⚾.

The 🎩 need **5** more 🧤.

Using this page: Have students compare the two sets of objects, then write the numeral that shows how many more are needed to make both sets equal.
Concept: **Comparing sets and finding the difference.**

138 · 6-3 More and Less

Count and write the number.

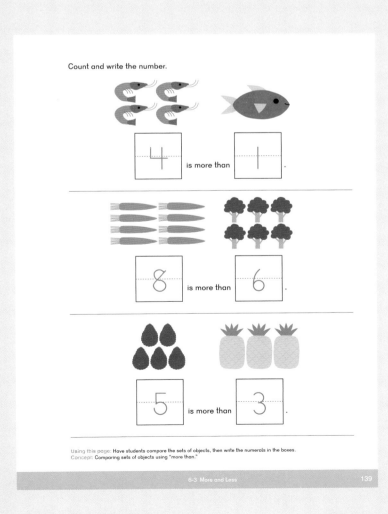

4 is more than **1**.

8 is more than **6**.

5 is more than **3**.

Using this page: Have students compare the sets of objects, then write the numerals in the boxes.
Concept: **Comparing sets of objects using "more than."**

6-3 More and Less · 139

Count and write the number.

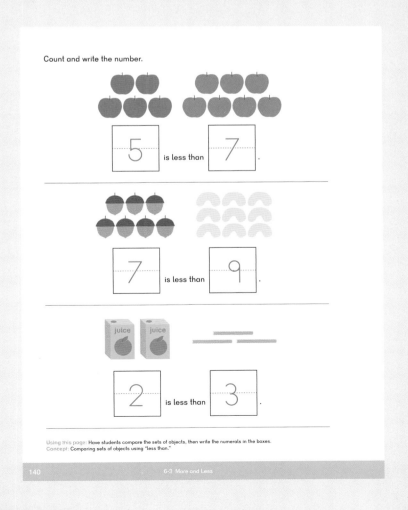

5 is less than **7**.

7 is less than **9**.

2 is less than **3**.

Using this page: Have students compare the sets of objects, then write the numerals in the boxes.
Concept: **Comparing sets of objects using "less than."**

140 · 6-3 More and Less

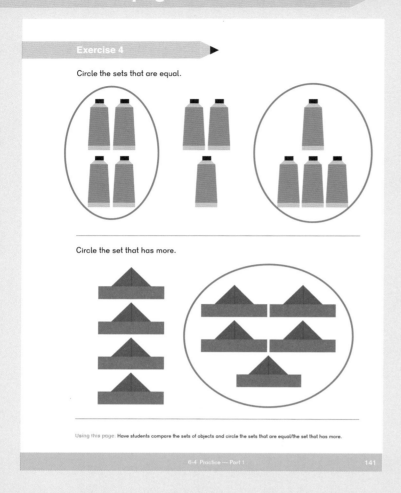

Exercise 4

Circle the sets that are equal.

Circle the set that has more.

Using this page: Have students compare the sets of objects and circle the sets that are equal/the set that has more.

6-4 Practice — Part 1   141

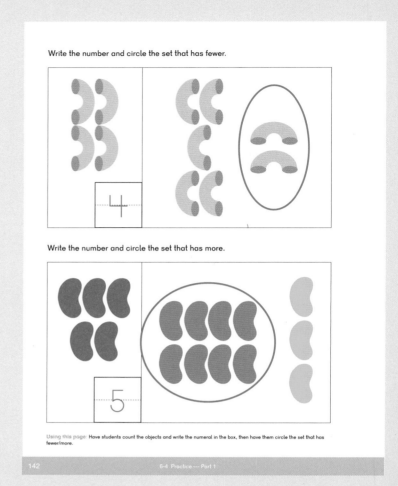

Write the number and circle the set that has fewer.

4

Write the number and circle the set that has more.

5

Using this page: Have students count the objects and write the numeral in the box, then have them circle the set that has fewer/more.

142   6-4 Practice — Part 1

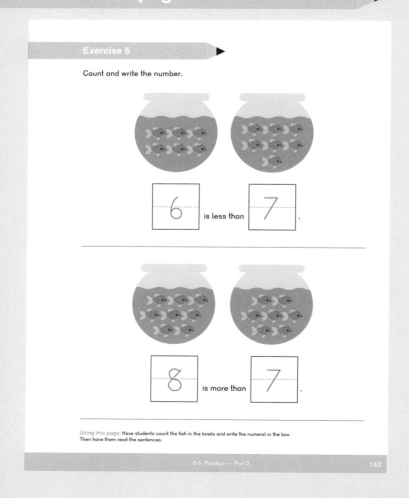

Exercise 5

Count and write the number.

6 is less than 7 .

8 is more than 7 .

Using this page: Have students count the fish in the bowls and write the numeral in the box. Then have them read the sentences.

6-5 Practice — Part 2   143

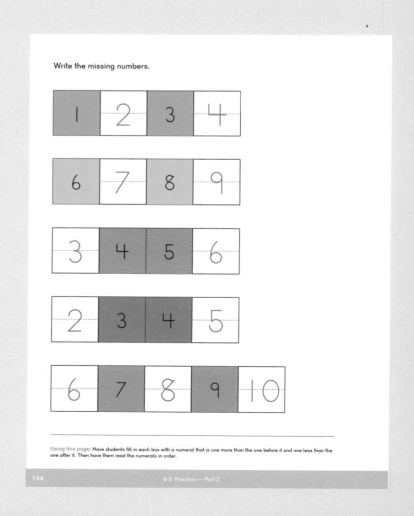

Write the missing numbers.

1	2	3	4

6	7	8	9

3	4	5	6

2	3	4	5

6	7	8	9	10

Using this page: Have students fill in each box with a numeral that is one more than the one before it and one less than the one after it. Then have them read the numerals in order.

144   6-5 Practice — Part 2

Blackline Masters for KA

All Blackline Masters used in the guide can be downloaded from dimensionsmath.com.
This lists BLMs used in the **Think** and **Learn** sections.
BLMs used in **Activities** are included in the Materials list within each chapter.

10 Dots	Chapter 3: Chapter Opener
Blank Five-frame	Chapter 2: Lesson 2
Blank Graph	Chapter 2: Lesson 11
Blank Ten-frame	Chapter 3: Lesson 2, Lesson 3
Five-frame Cards	Chapter 2: Lesson 3, Lesson 4, Lesson 5, Lesson 10
Fun Font Number Cards	Chapter 2: Lesson 3, Lesson 4 Chapter 3: Lesson 6
Number Cards	Chapter 2: Lesson 3, Lesson 4, Lesson 5, Lesson 10 Chapter 3: Lesson 6
Picture Cards	Chapter 2: Lesson 3, Lesson 4, Lesson 5 Chapter 3: Lesson 3
Single-line Ten-frame	Chapter 6: Lesson 1, Lesson 2
Ten-frame Cards	Chapter 3: Lesson 3, Lesson 5, Lesson 6
Vegetable Cards	Chapter 3: Lesson 6

Notes

Teacher's Guide KA